# TALKING THROUGH MY HARD HAT

*Chronicle of a 20th-Century Civil Engineer*

JOHN J. SCHEUREN, JR.

ZEITGEIST WEST

Published in the United States by Zeitgeist West Publishing
Visit us on the web: www.zeitgeistwest.com

Edited by Margot R. Scheuren and Claire E. Scheuren
Photographs used by permission from the Estate of John J. Scheuren, Jr.
Printed in the United States of America

Paperback ISBN: 978-0-9796921-7-8
Hardback ISBN: 978-0-9796921-8-5
Library of Congress Catalogue: 2026916835



Terminology and place names generally reflect the usage found in Scheuren's original manuscript. In a limited number of cases, editors have adopted contemporary style conventions or provided explanatory notes to assist modern readers.

"We civil engineers do our work and the rest of the world goes its own way—supplied with power, water, transport—by the structures that have been set up by the construction people. I can say that whether my works were in the open, where it was visible, or a buried pipe or tunnel, I am proud of having scratched a mark, however small, on a piece of the globe."

—John J. Scheuren, Jr.

♦♦♦

# Contents

♦ ♦ ♦

# Prologue

*Talking Through My Hard Hat: Chronicle of a 20th-Century Civil Engineer* is the work autobiography of our father, John J. Scheuren, Jr., an accomplished American civil engineer, explorer, and permafrostologist who conducted significant global construction projects between 1922 and 1986. During his lifetime, Scheuren received extensive honors and accolades for his work, including the Patriotic Civilian Award from the U.S. Army for his work in the national defense, membership in the Explorer's Club, and the naming of a river in his honor in Antarctica.[1]

Scheuren's autobiography, completed just before his death in 1995, provides character vignettes and salient historical perspective on engineering infrastructure projects, places, people, and societies across the world. His acute and substantive observations show Scheuren's shrewd mind, philosophical insight, moral clarity, and wry wit.

*Talking Through My Hard* Hat originated as a series of recorded reminiscences dictated by Scheuren during the final years of his life. Margot Scheuren transcribed the audio tapes, producing a manuscript of approximately 800 pages. Our father subsequently reviewed the manuscript in detail, correcting place names, technical terminology, dates, and other transcription errors. Because of the extraordinary scope of the original manuscript, not all material could be included in this volume.

---

[1] Scheuren Stream: A meltwater stream 1 nautical mile (1.9 km; 1.2 mi) west of Gneiss Point on the coast of Victoria Land. It issues from the front of Wilson Piedmont Glacier and drains northward to the Bay of Sails.

Above all, *Talking Through My Hard Hat* remains our father's firsthand account of his remarkable life and career.

The book captures not only Scheuren's experiences, but also the spirit of the century through which he lived—a life celebrated by M. R. Montgomery of *The Boston Globe* on May 20, 1995:

*Cape Cod Canal, 1936*

> *Like so many civil engineers of his era (he graduated from MIT in 1931), he could build anything—gas pipelines, dams, sewage water systems, airfields, and high rises. He could sink caissons and clear minefields. … Engineers of his generation had enormous practical abilities, often taking direct charge of machinery and explosives that would now be in the hands of a specialist. … (e.g., widening of the Cape Cod Canal in 1936) …*
>
> *The first plan for widening the canal, he said, would have required enough pumps to drain the Atlantic, and so he broke the job into management bites, shorter sections, and widened and rip rapped the canal behind sheet piling. When the day came to blast out the foundation of the old Bourne Bridge, located directly under the main high-tension electrical line feeding Cape Cod, he placed*

*the charges himself, and remembered with satisfaction that not only did electricity keep flowing, but the whole excavation slid neatly and harmlessly into the Canal.*

This book includes a rich fount of anecdotes, many involving historic and leading figures and events in Arctic and Antarctic exploration, Trinidad, Thailand, Egypt, the U.S., and Jamaica, plus in North Africa and Italy during WWII.

*Antarctica with Sir Edmund Hillary in January 1958*

*Talking Through My Hard Hat* includes descriptions of encounters with many other polar explorers and international public figures, including Sir Edmund Hillary, Bernt Balchen, Robert Peary, Jr., Admiral George Dufek, Maurice Ewing, Father Daniel Linehan, Governor Paul Dever, Martin Lang, and Sir Vivian "Bunny" Fuchs, as well as political figures in the U.S. and around the world, plus American construction giants.

*John J. Scheuren, Jr. with Robert Peary, Jr. and Ootah, who had been with Peary, Sr. at the North Pole, plus Ootah's grandchildren; c. 1952*

Scheuren's lifelong and extensive field notebooks are the source of numerous names and details that will be interesting and meaningful to students of history.

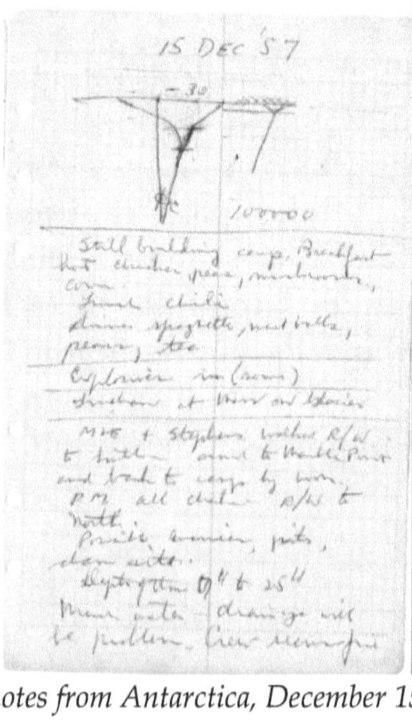


*Field notes from Antarctica, December 15, 1957*

Throughout history, civil engineers have made an enormous impact on the progress and development of the world. Although engineering achievements are recognized, the ideas and opinions of engineers are rarely noted.

We believe this memoir will appeal to all readers of history and anyone who has ever been curious about the wide-ranging issues and people behind: mid-century polar exploration; Cold War defense systems; and the construction of bridges, pipelines, dams, roads, tunnels, defense systems, air bases, and water systems.

John J. Scheuren, Jr. was a prime mover during the time when American infrastructure, from bridges to water supplies and defense systems, was at its very best. Currently, there is ample documentation that infrastructure is dangerously deteriorated. The history and perspective of civil engineers like our father is thus more than fascinating. Those values need to be actively honored and effected as part of a renewed and necessary focus on rebuilding infrastructure in the U.S. and around the world.

Claire E. Scheuren
Frederick J. "Fritz" Scheuren
J. Mark Scheuren
Margot R. Scheuren
August 2026

Editor's Note: Place names, project names, military designations, and technical terminology are generally reproduced as they appeared in John J. Scheuren's field notes, correspondence, reports, and recollections. During the period covered by this memoir, Greenlandic place names were often transliterated inconsistently by English-speaking engineers, Danish cartographers, and U.S. military personnel, and variant spellings frequently appeared in contemporary records. Where possible, modern or official spellings have been used; however, certain historical spellings have been retained to preserve the authenticity of the original sources.

♦ ♦ ♦

## A Daughter's Perspective

*John J. Scheuren, Jr. left a legacy that extended far beyond the structures he helped build, into the character of those who worked with him, and especially into the life of the daughter who came to respect him through the lived experience of working alongside him.*

*The following recollection of his daughter, Margot R. Scheuren, offers an intimate view into his world.*

The lessons in effective leadership, pride of ownership, ethics, and always striving for excellence in everything I do are what I learned by working for my father. He understood that the key to satisfaction was being valued and respected for the job you were doing, whether cleaning latrines or building a bridge. His employees gave him their best because they knew they were valued, respected, and recognized.

As a young woman in her early twenties in 1973, I was sure I knew it all: my father was an autocratic jerk, trying to tell me who I should be. For the first ten years of my life, he was a stranger, being in Greenland all that time. I saw my mother suddenly not in charge when he came home on infrequent trips from the north, with him telling me what to do. As a teenager rebelling against everything in the 1960s, my father was a pitiful example of establishment and rule. If he said it… it must be wrong.

As we were protesting against the Vietnam War and for civil rights, he seemed to be part of what needed changing in the world. I really believed we'd make change, if not for our politicians and our "elders," who thought we were wasting our time. His ideal was for his sons go to college and, hopefully, follow his footsteps to engineering professions. His daughters' educations were considered important but not necessary for their future. He seemed hopelessly part of a generation not moved to change the status quo.

He was authoritarian to us and unable to communicate to me or my siblings on any level. He was lacking in affection. We had to accept the fact that he could not be a loving, supportive father unless we were doing something he valued.

He used to say he was a "performance snob," and I came to understand this meant performance of any kind—any job, how you treated people, or how you held a shovel. In 1973, somewhat floundering in purpose, while I was trying to "find" myself, I found myself living with my parents in Long Island, New York. My father offered me a clerical job at the engineering project he was currently running. Being between jobs and states (I had been living in Boston for work and moved to Connecticut to be with my boyfriend), I decided it would be something to try, as I did need a job.

I got an instruction book on typing and spent a few weeks learning it. My father introduced me to a City of New York employee who was supervising the work for the city, and he agreed it would be okay for me to take the job. My father also had to work through his employer's nepotism rules to hire me. He was showing faith in me, and I began to see why he wanted to help me in this way. He actually was mostly hoping to get a chauffeur.

We lived in Long Island, and the project was in Queens (Rockaway Beach). He was Resident Engineer of a very large sewage treatment facility upgrade–a resident inspection force of about ten engineers working in a trailer on the job site. So, my job as clerk included a lot of driving, in addition to chauffeuring him back and forth to work every day. I had multiple trips per week to downtown Manhattan, airport trips to LaGuardia and Kennedy to chauffeur the many other engineers coming for day trips from our Boston office, trips to Brooklyn to deliver paint samples, and meetings in Manhattan and Queens with city and federal people overseeing this project.

Spending every day with my father (whom I shall call John in this narrative, as that was what I call him now), alone in the car traversing the many highways from Long Island to Rockaway, in rain, sleet, ice and snow and during the gas crisis, we developed a relationship built on which road to take, while listening to news radio. At work, we hardly interacted at all – I was to answer phones, type reports and drive.

The main crew of inspectors under him included an Assistant Resident Engineer, Desmond Smith. He had hired Desmond in Trinidad and respected and trusted him. John was Desmond's mentor in work and life. He obtained a work permit for him to move to the U.S. to take this job. Desmond was a mixed-race (Black-Chinese) Trinidadian and excellent engineer. He had the full trust of John and went about his job knowing John had his back.

As I mentioned, this was 1973, and the ranks of the construction staff were almost 100% white males, the only exception being the Native American iron workers. Another staff engineer was Marilyn Roberti, the only other woman on the site. She was the "civil" inspector, thus inspecting concrete, roofing, steel, etc.

We had an inspector for each discipline: electrical, mechanical, civil, structures, plumbing, HVAC, and controls etc. We had an inspector for the outfall work (a large underwater pipe), a former longshoreman named Harry Corfield with whom John had worked on previous pipeline projects.

There were five "prime" contractors on site, with a total of over 300 workers. The largest contractor was "structures and equipment," with Don Anspacher as their superintendent, with a trailer next to ours. As you might imagine, the "specifications" for this project were volumes and volumes, covering every facet of the construction–buildings, pipe galleries, treatment tanks with their systems of electric, water, heating, ventilating, controls–all in intricate detail. John kept these volumes on Desmond's desktop to the left of him.

There were also the drawings for every detail; these were kept in another room in the trailer, where we had a small kitchen and the rest of the team had their desks. They were laid out on tables, ready for checking and review as each step in the project progressed, as were the specs. John was ready to show the contractors what was what on every visit.

When Don or other superintendents came to him to question something, he would say "Well, section such and such of volume such and such says you will do it this way...," and Desmond would flip the volume open to the section and put it on John's desk for the contractor to see. It seemed he had a photographic memory of the entire project specs, and it was quite intimidating to all, including the inspection staff.

As John notes quite extensively in his autobiography of construction work, he was "following the fake" of the profession of construction with his expertise as engineer. What I came to understand was making sure the job was done to the best professional standards. Stop the cheating being prime standard: no bribes, and no coercion or skimping on materials or substandard workmanship.

Metcalf & Eddy's engineers had drawn up all the specs and drawings for this massive project. But what a layman might not understand is that converting from words and pictures to build an actual working treatment plant of this magnitude is not straightforward. The Resident Engineer and his staff were there to make sure it was a safe, working facility and that the contractors used the right materials and method of construction to make it so.

Often, we had conflicts between what was on the drawings to what was reality in the field. We had many visits from the design engineers in Boston to resolve these conflicts. The contractors learned very early on that John and his group were not going to be pushovers or subject to bribes to overlook something.

John wore baggy pants that hung over open laced boots, and he went everywhere on site, looking at everything. They thought he was an old man trundling around. Not the sharp-eyed fellow they soon found him to be.

Many efforts were made to overrule his inspectors. One such incident involved Marilyn. She had rejected a concrete load because the truck had idled too long, and she felt the concrete was affected, so she'd ordered the truck sent back to the plant for a fresh load. The contractor came storming into the trailer, already yelling before he even reached John's desk. He ranted and raved about the return and demanded the truck be allowed to stay and deliver the load.

When he finally stopped ranting, John calmly asked, "Did she tell you to return the load?"

That started the ranting all over again, and when he finally ran down, John said, "Send the load back, and stop wasting my time."

Later that day, when Marilyn came back to the trailer, John talked to her about the load. He said he thought that even though, "technically" according to the specs, the load should be sent back, other factors could be considered, and she could have accepted the load based on those factors. When I asked him on the way home why he'd handled it that way, he said, what would be her value on the job if he'd overruled her decision?

That was the first of many times where I saw him back up his staff's decision-making, and I began to understand how critical it was. This happened with all the inspectors on site, but particularly to Desmond and Marilyn, because of the distrust of having a mixed-race man and a woman making decisions.

Bribes and favors as a standard way of doing business in the construction world became very evident early on. I didn't always recognize them, but of course John always did. A gift dropped off was always returned with thanks. His discussions with all staff about taking anything—tickets to a game, a bottle of booze, or any favors—would be strictly enforced with immediate firing. He allowed staff to accept a bottle at Christmas, but he himself never did.

The largest contractor on site had hired an "expeditor" to help with getting building permits. He appeared to be a slick operator, used to working with New York City to get paperwork through. We frequently went to visit him at his home office, to pick up permits or documents on the way to or from home.

The first visit we made to him, he asked that I stay in the car, but John insisted that I be part of the meeting. ("She can hear whatever you have to say to me.") We sat in his office, and they discussed many aspects of the permitting and the project. During their conversation, he passed an envelope across the desk to John. It was not talked about. John picked it up and looked inside and then passed it back across the desk.

Back in the car, I asked what that was all about. He said he was offered a bribe to help the contractor. I asked, if he knew what it was, why did he touch it or open it?

He replied, "Because I wanted to see what he thought I was worth." John always said that every man has his price. He said that his was his wife and children.

His integrity, and his employees' integrity, was put to the test every day on this project, as he had experienced on every other project he worked on in his career. Not just with outright bribes but coercion, as well.

In addition, there was the daily work of making sure the correct method and materials were used. The roofers called Marilyn to inspect their work on one of the buildings. They had applied a coat of elastomeric roofing to one of the buildings. On display were many 50-gallon drums of the material used.

Based on a calculation of the size of the roof and how many layers were specified, she knew how many drums they would use for the first coat. They didn't notice her make a chalk mark on each barrel. They called her back to inspect the next layer a few days later. The drums were the same ones. She pointed that out and told them to bring in new material and get the job done right. Most of the attempted cheating was in materials like this.

We had to bring samples of the paint being used in a pipe gallery, because the specs at that time called for a certain concentration of lead to protect the pipes from rusting. The inspectors weighed the paint cans as they were being applied, and when it looked like they were light, we sent them to the lab for verification. In those cases, the contractor had to redo the painting.

One of the biggest "rejections" of work came from John. There was a very large gas holding tank at the site, right next to a train line. The specs called for very specific welding techniques to be used to protect the integrity of the tank. This work was to be verified by X-raying the welds. In addition, they were required to submit specific documents for the welder.

When the contractor could not produce the X-rays for many of the welds, John rejected paying for this installation until they redid the welds and provided X-rays of the new welds. They pushed him every which way to approve it. He refused.

They contacted the higher-ups at Metcalf & Eddy to convince him to approve it. He refused. The city pushed him to approve it, and he would not. He was determined that he, his company, and the city should not let this stand. If there were a spark from a train and it ignited this tank, it would be a catastrophe and the blame laid on those who'd approved this, rather than the perhaps substandard work by the contractor.

They produced certificates of the welders' work and experience but would not agree to redo the welds that were not X-rayed. The city eventually accepted the work and paid the contractor, but John's signature or signature of an employee of Metcalf & Eddy was never provided.

We can all experience this type of issue on road construction. We wonder why the roads deteriorate so soon after new roads are paved. Mostly, this can be traced to using less or substandard materials during the construction. These projects are awarded to "low" bidders.

Projects go out to bid, and contractors are supposed to submit a plan for work: people, materials, methods of construction, etc. that meet the drawings and specifications for that job. But contractors are not selected based on that; rather, by the cost. They show, for example, they will use so many yards of concrete of a certain strength and the prices are already much known by the designers. But when the actual project is won, they proceed to use less material or cheaper material to make their profit. The contractor who fairly costed the job is not the one who wins it.

John always stressed to us that we should work hard at whatever job we had. He said the reward was pride and sense of purpose. There is pride in being the best at sorting photos (which was one of my summer jobs). Regardless of the importance of the job, the point was to give it your best and to work to achieve excellence.

Bricklayers are as important as the architect in creating a structure. There is no place for superiority based on education or wealth or of race or circumstance of birth. John practiced what he preached. He hired people whom he judged as hard workers–knowledgeable about their area, but open to listening to other points of view.

He always listened to his staff–really listened–before making judgments. I witnessed the hiring and the firing. He had no tolerance for someone who did not have pride of ownership for their work. Dishonesty in work was where he judged people. He was unmoved by those who acted superior or above anyone else. One engineer was fired for his mistreatment of subordinates; another for not speaking up when seeing a problem.

Leadership, I believe, is not easy to teach. It has to be by example to understand its failure or success. Picking the wrong person for the wrong job is bound to cause frustration and angst for both the worker and leader and, of course, results in less than excellence in the work.

When he was in charge of the project in Greenland, hiring was very key. The conditions of the Arctic and the distance from home were a great equalizer. The cost of getting a worker moved up to Greenland and the cost of sending them home were of great importance–not to mention their effect on the morale of the larger group. The main reason for his trips home were mostly for candidate interviews.

I use the lessons he taught me every day, in my work and home life. My husband and I have told many "Pappy" stories to our son and, indeed, to many other friends and co-workers.

I was fortunate to work with him and, from the correspondence he left behind, I can see that many others felt the same way. I will leave off with a very funny moment during this project.

As Resident Engineer, John had a parking spot near the trailer with a big sign "Resident Engineer Parking Only" (probably 3'x4'). As I was in and out many times during the day, it was important we had that parking space.

The sheet metal workers were known as an obnoxious bunch, refusing to wear hard hats as they were much too macho for that. One of these fellows loved to park in that spot and was told off several times.

One day, John had me buy large stickers—8-1/2" x 11"—in bright orange. When the worker parked there again, he told me to go out a paste the sticker on the driver side front windshield.

At the end of the day, the worker came screaming into the trailer, raving and swearing about not being able to peel it off and how dare we. John just looked at him and, when he finally stopped swearing, calmly asked him if he saw the sign for Resident Engineer parking only. The fellow said, no, he hadn't seen the sign.

John replied, "It's a good f**king thing it didn't say Land Mines!"

♦ ♦ ♦

# Introduction

Some people have recommended *"Talking Through My Hard Hat"* as an apt title for my story.

"Hard hat" seems to have developed as a name for a hard-boiled construction worker who was generally not exactly liberal and supposed to be more likely to vote for somebody like a Nixon than a Humphrey. This does not apply to me or most of my friends, however.

Hard hats are construction stiffs. *Stiffs* comes from the term working stiffs, who were sometimes itinerant workers, perhaps hobos. Construction stiffs may be trained workers, tunnel workers, hard-rock miners, or soft-ground tunnel workers, who are usually called sand hogs, and maybe pipeliners. Although pipeliners are always called pipeliners and seldom do other types of work, a construction stiff is usually a generic term.

Hard hats are all of us who follow the work of construction as a profession or trade or, as the old pipeliners used to say, "follow the fake." Construction stiffs are really people like Louis Perini, Glen Maxon, Charles Foreman, Bert Stuart, Warren Shoemaker, and Bill Naumann. All of these presences ran large construction contracting firms.

And they are people like Joe Hachey, the best carpenter foreman I've ever met; like Dirty Face Bill Jones, a hard-rock tunnel man; Andy Paton, who had been both Chief Engineer for an engineering firm (Metcalf & Eddy) and Superintendent for General Contractors; Mark Klinefelter, master diver, iron worker, superintendent and manager for both contractors and engineers;

Cherokee Marley, the best pipelining foreman who ever bossed a pipeline spread; Randall, skilled amateur magician, who could really build bridges; Keith Wason, Kiewit Company General Superintendent, with his expression "who's going to see it," who got all sorts of work done at Thule, Greenland and other places in the world; and like Bill Smith, a British soft-ground specialist and great tunnel man.

My part in "following the fake" or trade or profession was on five of the six continents, all except Australia but including Antarctica. We would get the job done on the ice caps, both north and south, and on the deserts, both cold and hot. Through the mountains and under and over the rivers, across open country and into the teeming cities, each one of us developing certain specific skills and expertise. But we also developed a paranoiac (perhaps) way of thinking that all people outside the construction business are *civilians*. The Italians say, *borgese*–non-professionals.

Sciences and professions, which we all are close to and part of their general scheme, seem to use the word *layman* for any non-scientist or non-professional. But like many of my coworkers, I think of non-construction stiffs as civilians; so, there I have a pride and affection for "mine own people," to borrow a phrase of Kipling's, and I will try to describe their actions and reactions to the world we all live in as construction stiffs and civilians.

As I continue in this book, I shall share many brief vignettes and notes on people I have met over the years, 1922 to 1986, plus my thoughts on the places and societies to which I have been exposed.

I realize, as I record the chronology of my working life, I may have dwelt too much on the struggle, the pathetic, and the pitfalls, in line to the various systems of culture, politics, and human frailty, while slipping over the satisfactions I observed in others.

Lincoln wrote that he is proudest of the approval and esteem of his fellows, the men who understand exactly what he is trying to do in this narrow world and are competent to judge his doing of it. To quote Herbert Hoover, "The only accolade the engineer wants is the approval of his fellow professionals."

Col. Morton Solomon expressed his thanks to me for writing in praise of his accomplishments in Thule, Greenland. I said that he had more citations than many people with higher rank, and I had commended him. He said, "your opinion is most valuable, you really know the story."

Father Dan Linehan wrote to Mr. Eddy (Metcalf & Eddy) concerning our Antarctic work and said he really appreciated the opportunity to work with me. He said Jack Scheuren and the M&E (Metcalf & Eddy) party were excellent people and that whereas, with an exhibition of scientists, things were confused, if that were these engineers, it was always planned and well executed.

Dr. Robert L. Nichols, who was on the same expedition (Marble Point in Antarctica) wrote to me, "Jack, all those many years you have been a leader and a good one, too, and an excellent organizer and a well trained and experienced engineer. All this and much more was shown at Marble Point."

I have Dave Krim's memo on my work at Rockaway, New York, and General Wilson's citation on my work in Greenland and the Arctic. The Chief of Engineers, Lt. General Weary Wilson, said I exhibited a high degree of enthusiasm in solving the many problems encountered, and my accomplishments reflected great credit to the Corps of Engineers and the U.S. Army in an able accord to carry out its military construction in the National Defense.

And Andy Paton, on the occasion of one of my retirement parties, said, "You are the most scholarly construction stiff I ever knew." I shall write more about Andy Paton later.

Of course, the approbation of my fellow professionals was shown in the several times after my first retirement in 1971, my being called back by Perini in '72, by M&E in '73, again by M&E in '76 and 1979-80, plus for a few little consulting jobs by Camp Dresser & McKee (CDM) in 1982 and 1983-84. All of these have to make me proud, when I see some of my old field notes. I wonder how some of the projects turned out so well, despite all the problems of the people, the clients, and the politicians, etc. But we always got it done. Deferred completions were in Alexandria,

Egypt and Kingston, Jamaica. But in both these places, I am sure, in time, though perhaps not in my lifetime, some of the groundwork I laid down may be built upon to bring success.

In 1956, when I was on the *Westwind* icebreaker, we had to split parties on a portion of the investigation. I remember Captain Tach, USAF Engineers, reporting to me on his part of the survey. The going was so tough on one portion of the route that, "Even you," he said, "might have hesitated to cross."

In 1976, when I was in Bangkok, I got a letter from Harry Corfield, who had been my pile-driving man at Rockaway and was then driving piles in Arabia. Harry said, "I hope the mess you have to straighten out in Bangkok is not so bad that even you cannot succeed."

I have been with others whose pride in their work shone through. When Sir Edmund Hillary visited our camp in Antarctica, he was impressed with the heavy equipment we had working on the edge of the glacier. He was climbing another glacier, and this particular one in Antarctic was easy. On another one in the Himalayas, he was cutting steps on the ice with his ice pick to climb. His eyes were shining as he described all the chopping and climbing.

Maurice Ewing, when he visited our work in Trinidad, spoke with us on the use of low-value land to treat sewage by means of stabilization ponds. We talked of the depths of selenite crystals in our core borings. He was interested in that, but then his whole face glowed when he told me about an advanced type of sampling device that he had made to sample sea bottom sediments.

I have memories of Charlie Ractriner, a mason foreman, stepping back to look lovingly at the band of colors and stone sides he had worked into the sea face granite on the wing wall of a bridge. Joe Hatchey, the great carpenter foreman, shaped his forms to produce details and finish on a concrete bridge (now destroyed to build a larger one), and that bridge went on to win a record award for excellence in bridge construction.

I have heard Cherokee Marley, pipe foreman and Henry Fall Leaf, a superior welder, go over a difficult piping layout in a compressor station for natural gas in Kansas and being proud of what their decision was, and then watching the thing come to completion. Baldy Hoyt, I heard him speak once of the pipe crossing the rocky stream: "Look at that—it swings, loose as a goose." There are many tales. I could go on for days.

When I blasted out the foundations of the old Bourne Bridge over the Cape Cod Canal, there were a dozen small footings on shore and a larger abutment at the water's edge. High-tension lines were overhead, and the explosives had to be carefully set, so the smaller piers were broken up into smaller pieces for our scrapers to pick up and take to the fill. The abutment had to be blasted so the *Crest*, the dredge, could scoop them up as she traveled. Even after over fifty years, I remember how the General Superintendent, Jack Doherty, arrived at the site and praised me for the excellent manner in which everything fell into place without hitting the power lines and without slowing the dredge or our scrapers.

When I was in Italy during the war, I had just been transferred to an aviation engineer battalion that had been overseas for three years, but whose state of training was still poor. They could blast out a pit for a latrine and laundry, and there were only two blasting caps and fifty holes to be fired. There was plenty of prima cord. I loaded each hole and completely and carefully tied them by the cord hooked to the blasting cap on the end of the cord, and then fired the shot.

I was pleased to hear the exclamations of the men, "Oh, it really went off." They had not been trained how to use it.

When I left Jamaica, because I had become close to the dedicated but frustrated thirty-plus-or-minus troops; superintendents, plant operators, and chlorination and laboratory people, I went to a farewell luncheon, where they presented me with a Jamaican aluminum tray, some locally made tablecloths and napkins, and samples of miniatures of all the liquors made in Jamaica.

I paid for the lunch at the Yacht Club. These men paid $10 apiece for their presents, and they didn't get paid very much, but I was impressed by the eulogies they heaped upon me. One of the superintendents came to me afterward and said he didn't make a speech, because he was too emotionally involved, too emotionally moved.

Since then, I have received letters from some of these people. One of them asked, "What did we do that you left us?" Others said they had appreciated what I had done, which, because the director itself was not really much, they hoped the blueprints I had set up for the management would be implemented, so the National Water Commission would get out of the hole it was in.

Pride to work. These people would haul pipe in their own little cars because of no transport. Rudolph Brown performed just short of miracles to keep the chlorination going using Rube Goldberg-type setups. Anyway, despite the lethargy of the director, I feel proud that I got to these people in the lower ranks, who were really close to important work.

I had a conversation with Bill Nauman, who was the Chairman of MM Sundt in Arizona. He is a month older than I, but he had been working a lot longer (Bill had no college degree). We talked of the days when we were little boys, handling horses and learning to run equipment. He was blacksmithing, and we both believed that being in construction is a satisfying experience.

As Hoover said, "The greatest liability of the engineer is that his work is out in the open, and if his works do not work, he is damned."

But I can say, whether my work was in the open where it was visible or a buried pipe or tunnel, I am proud of having scratched a mark, however small, on a piece of the globe. Along with my fellows, I am able to remember my own feelings of pride. I believe our greatest memories from sixty years of construction are those of the people we work with.

As I mentioned earlier, some of us stiffs consider the world to be divided between construction stiffs and civilians; it is probably difficult for a construction stiff and a non-stiff to understand each other. But perhaps any understanding is not necessary. We civil engineers do our work while the rest of the world goes its own way; civilians have been supplied with power, water, transport, and so forth by the structures set up by the construction people. Of course, the civilians, even if they are scientists (a term created long before we had engineers), often complain of the hardships and terrors induced by technology. Sometimes, they think an engineer can do anything.

Bernt Balchen proposed that the engineer could build a highway across the Greenland ice cap. Paul-Émile Victor proposed that engineers should organize tow boats to haul icebergs from Antarctica to Arabia. The questions of real value of a road across the ice cap or the feasibility of hauling icebergs at all may be answered negative.

I pointed out that my friend Balchen was not a civilian sent to a construction site. He was a great explorer and commander. Col. Balchen had a feeling that engineers could do wonders.

Dr. Bader, of course, realized that the engineering challenge in moving these icebergs would involve his particular expertise in the world of snow and ice. He, therefore, wished to charge the engineers, who knew about tow boats and the problems of horsepower and handling ungainly tows in the storm-ridden seas surrounding Antarctica. He really wished to drop the problem on the engineer's lap.

Just before I went overseas in WWII in the U.S. Army Corps of Engineers, my wife's late brother, who was at the time an editor of *Fortune Magazine,* told me that technicians were being sent to straighten out the world's problems. He perhaps should have been more skeptical of the engineer's ability and the probable interference by political roadblocks.

But it still is a bit stimulating to have non-engineers have such simple beliefs about the worth of engineers and construction stiffs. Of course, most civilians also find fault with the use of technology, so we are damned if we do and damned if we don't.

# CHAPTER 1

# 20th-Century Construction Notables

## Louis R. Perini

I first met Louis R. Perini (LRP) in McNamara's Transit Mix Concrete plant in Allston, Massachusetts. I left MIT to work for a year and save my money, so I could return my senior year as a gentleman. No cleaning floors, toilets, punching adding machines in the bursar's office, lumping in the foundry, or working as a laborer. I worked on construction and as a bouncer in a dance hall.

I was back at school but doing a bit of work for McNamara on testing aggregate with concrete. Louie Perini was planning to pave several sections of the Boston-Worcester Turnpike by means of transit mix concrete, so there were tests to be made to satisfy the State Highway Department. McNamara had asked me to operate the plant for these tests, so I came up from school.

Mr. Perini watched me very carefully and then asked me what my future plans were. I said I was going to graduate in June 1931 and would be looking for a job. There were very few at the time. He said, "When you get out, I'll give you a job." And that was it.

When I graduated, I had my tonsils out at the MIT infirmary. Then, I traveled to Ashland, where Perini's office was then, and LRP was out. Ted Guiney, the truck mechanic, whose brother Jack had a truck working for Perini, had a truck working for J.H. McNamara. I stayed with Ted and his wife and traveled with him and his van and the transit mixers for two to three weeks.

Finally, I met Louie, and he said he knew I was on the job and was glad to catch up with me. He gave me a stopwatch and a notebook and told me how much he could pay. He said he wanted me to analyze his concrete operations. I did and reported to him that one of the plants had a soft spot on this job.

The superintendent, McCarthy, did not welcome me the way the others did. I pointed out to him that his trucks made an extra stop to take on water that the other plants didn't. He gave me a bad time, so I went away. Then, Louie asked me why we seemed to be getting a bit behind on that plant. I told him, and he told me to fix it. I got a truck and a couple of laborers, bought some pipe and fittings, and straightened the matter out.

Louie was pleased but McCarthy became my enemy, though not a dangerous one. After a season or so on the road building, I worked in the office on the estimating. The next spring, Louie brought me out of the office and showed me a picture of a transit mixer he had developed that had a side discharge. This made it valuable for highway slab work. Then, he sent me to the Quabbin Reservoir for tunnels and caissons. After that, I went to New York City for the Grand Central Parkway work.

All this time, Louie was a calm, teaching person. He was only eight or nine years older than I. His father had died about seven or eight years before. Louie had four brothers. The oldest, Frank, had argued with Bonfiglio, the father, then left and never returned. The next oldest was Fred, who was a fine, friendly man, but Louie was the one who took over. He was the operator; Joe was a year older but became treasurer. Charlie was my age and still in school.

They used to call Louis the Duke; he had a small mustache in those days. He once said, "Why do they call me the Duke? I don't put on any airs."

Which was true, and he had a mind like a filing system—today it would be like computer memory. He was interested in baseball and finally owned a ball club. He could be talking expertly about the amount of cover needed for driving tunnel, next discuss the type of stone facing required for a new bridge abutment, and then shift to a statement on the pennant-winning chances of the Boston Braves, all in a few minutes.

When, at twenty-three years of age, I went out to my first job alone as a general superintendent on a New Hampshire highway bridge, he gave me this advice: "The only thing I want is to get paid for what do you, keep track of what you do. If you do ten cubic yards of rock, don't try to get paid for twelve, and you damn sure don't settle for eight."

From 1931 to 1940, we worked on many kinds of projects. We started by subletting the bridges. One day, he asked me if he could transfer the bridge foreman I had (we were building that bridge ourselves). I said, "We'll save his pay. My foreman, Joe Hatchey, and Angelo Broseali, another foreman, we'll do the job." And we did.

He was upset about the Cape Cod Canal job where the other contractors were getting extra slips and cheating the government. He never wanted that kind of money. When he decided to build his own tunnels, we were struggling along. One day, he said, "Why do we have to learn everything the hard way?"

But we did, and we learned all the rest of us. When he wished to look at the pipeline work, I was sent to Kansas City and met C.S. Foreman, then I came back and resigned. Louis said, "That's okay. If you ever want to come back, let me know." He later asked if I wanted to come back on the big job in Pennsylvania, but I was going into the Army.

We kept in touch. Once, in 1956, I met him in Boston outside Bill Callahan's office for the Massachusetts Turnpike. He talked for a while. Usually, we had met at the Engineer's Club. He asked me if I was not getting sick of traveling around as overseas manager with Metcalf & Eddy (M&E).

I said, "Louis, when I worked for you, I was like a man on a checkerboard."

He laughed, and we parted. The next day, at lunch at the Engineer's Club, Charlie Richardson, the vice president of Perini

Marine Division, said, "You know, our manager on the pipeline in Canada dropped dead a day ago." Louis never missed a trick.

When he died, in 1972, I was back working for him, just leaving for New York City. I went to his wake. His wife, Florence, asked me what I was doing, and I said I was about half retired. I had been retired from M&E forcibly that year.

She said, "Retire completely. Don't be like Louis." He had all the children except Albert on the line then, there alongside the casket.

I said, "Florence, Louis is an old construction stiff, and old construction stiffs never quit."

This pleased her. She told all her family, "Do you hear what Mr. Scheuren said? Old construction stiffs can't quit." That was the last time I saw her or old Louis.

Albert, Louis's only son who was really a construction man, wrote me later that there could be only one president at Perini, and his brother, David Perini, was that one.

My last remembrance of Louis Perini was in Alexandria, Egypt, when Sami Abdelkawi, the chairman of our Egyptian partners, told me he thought Louis Perini was the greatest man in the world, and he wanted to write a book about him. I wrote to Albert about that, and he answered that he had been with his father when they were in Egypt, working on the job in the high dam, which the Russians got because Dulles didn't want to get into it. LRP had talked to me about that a year before, but it came to naught when Dulles cancelled it.

Albert said, "My only regret is that dad died too young." I agree.

## Charles Stanley Foreman (CSF)

We mostly called him *Boss*. None of his employees called him Charley. He was Mr. Foreman or Boss.

He was in his mid-fifties when we met. He had come from a boyhood on a farm in Ohio. He used to tell me how they bathed in a tub in the kitchen near the window, so the sun could heat the water, besides the heat from the adjacent stove.

Early on, he had worked in Kansas City for Public Works Sewer and Water, and then he started his own company for tunnels, pipelines, pumping stations, etc. At this time, his company had several jobs going on. They were mostly in the natural gas pipeline companies, oil companies, and some waterworks.

CSF took me out on the work in mid-Kansas. He had a couple of hundred miles of 26″ pipeline on that job and several river crossings, each one two lines of 20″ pipe. This was called Big Inch at that time. Any pipeline larger than 16″ or so—Big Inch, as the pipeliners said, if they had trouble with such pipe—the hole is too big.

CSF had a warehouse outside of North Kansas City, where his equipment was stored and repaired and shipped out as required. This was run by Chief Lightfoot West. He was a huge fellow. We got along fine. He told me stories about the times he had on some of the jobs in wide-open spaces with the camps and keeping order. Once, he said, he fought all night.

CSF was a different type from Louis Perini, but he was just as scrupulous and honest. After I had spent a few weeks with him, traveling over the pipelines and meeting his superintendents and foremen, he asked me to go to work for him. He said he would make me general superintendent. I protested that I had never seen a pipeline before. He replied that I had worked on roads, tunnels, bridges, canals, etc., and that the sole criterion was to get the job done. "Get it done" was his motto.

I went back and made my peace with Louis Perini and, as the pipeliners say, came back and set in.

CSF hated changes of plans. Once, I had figured to use well points on the river crossing based on some samples that tested out to be prime material suited to well points. However, the material under the river was not uniform, and the well points could not cool the water down, so I changed the plan and the method.

CSF said, "It was your own plan. You should have persevered." He disliked changing plans. But, of course, sometimes, you have to do it.

He used to say, no matter what the job was, it had to be done. One, get the job done. And two, get it done right. And three, make money off it. But it had to be done no matter what. If you were faced with a decision you had to make, make one, and you have a fifty-percent chance of being right if you made it, whatever you did. So, you went ahead, and if you were wrong, you adjusted it. As experience increased, you could cut down on the percentage. But if you stopped and did nothing, you would be one hundred percent wrong.

Pipelining, and this applies to main line work or river crossings, is a profession that requires initiative and imagination in devising expedient methods. Mr. Foreman really appreciated it when one handled a new job through an unusual situation and succeeded. Even if it cost more money than he had estimated, he was ready, like Louis Perini, to take the loss as quickly as possible and go onto other work, where he could make the money back.

When the war came along, we got some work on the War Emergency Pipeline in Texas to New York, also called Big Inch Line. We had several hundred miles of this main line: 24" and 20" were the two lines and thirteen of the major river crossings. This was our first cost-plus-fixed-fee job.

All the contractors, except one who was later investigated for possible sabotage, had their best crews and pushed hard, so the crews would not lose their techniques. This cost-plus-fixed-fee setup was that the cost was charged for all equipment, labor, materials, and so forth. The fee was at a rate of so many cents per foot, which covered our main office and my salary.

Because all the contractors finished ahead of time, with the actual cost below estimate for the emergency pipeline, they still renegotiated a fee after the war. The cost of this renegotiation gave us the tax benefits, and so, it was a waste of time. And while we had to give some money back for the fee, we got it all back in tax rebates, so actually, the government spent more money than it needed to. This was our first CSF job.

In Ohio one time, when Mr. Foreman was visiting, he and I had stopped by a road boring crew pushing the casing under the highway through which the pipe would be pulled. This crew was mostly local men we'd had to hire. The job was not going right.

I started to get out of the car to give instructions to the foreman. Mr. Foreman stopped me and said, "Don't bother. I've had all I can stand. Take me to the airport now. I'm leaving." He was pretty well shaken up then.

As we drove away, he said, "Well, I thought I'd never say such a thing, but it's cost-plus, anyway."

Some employers I heard were quite upset if their key men went into the Army. But not Mr. Foreman, he never said anything one way or the other. But when I told him I was going, he seemed very happy. "Always knew you would do it."

## Andrew C. Paton

Andy was resident engineer for the State Highway Department when I was sent by Perini on the Barber's Crossing job in Worcester. We had a good relationship and brought the job in in good shape.

He kept in touch over the years. He left the state and worked for M&E in Carolina, until he went in the Navy in the war. Then, he left and worked for Henry Lundgren for a few years.

In 1951, M&E got the Thule, Greenland work. Mr. Eddy, obviously a great judge of men, got Andy Paton to come to work for M&E to be in charge of the job. Andy wanted me to join him. I was building some bridges in the Berkshires. My wife was going to have a child in several months, which actually were twins; I wouldn't go to Greenland until several months after the twins were born.

The work increased, and we had more sites all over Greenland and Northern Canada. When Andy asked me to go, I demurred. I was a construction stiff not a design engineer. But Andy insisted that I was just what he was looking for, so I finally became M&E's

overseas manager. Paton ran the stateside office in New York and Boston, plus I got my chance to be an explorer in the Arctic and Antarctic and become knowledgeable about permafrost and ice caps, snow and ice in Antarctic and Arctic construction.

Andy Paton was a fine gentleman. We were close friends for over fifty years, from 1935 to 1986, when he died. His belief in me and my trust in him brought me many fine experiences.

He and I were both dumped from M&E in late 1971, and Andy never went back, probably because he was close to the people in the conglomerate, and they had bad feelings. Mr. Eddy and Andy completely parted from Research Cottrell. I was not known to those people at that time. All they knew was that I was sixty-two years old. Andy was three months older.

## John Baker

John Baker was resident engineer in the Pittsburg Dam in New Hampshire. Although he was not a young man, he had a young wife and three growing children. He was experienced and helpful.

He used to say, when you have to sit down and tie your shoe laces, you were either out of shape or too old.

Baker often stated, "No man is ever hurt on dangerous work." That has been shown many times. Because it looks dangerous, you will take care and observe proper precautions. On run-of-the-mill situations, one becomes careless and does not take proper precautions. Then, accidents occur.

On the job we were building, we had lumberjacks, scalers, and all sorts of timber men and rock cuts and so forth. When the insurance man predicted, based on the size of the job, that one or two men would be killed, I chased him off the lot with a pick handle. We only had two injuries, and neither one lost time.

The two non-lost time accidents we had were the time keeper, who jumped on the running board on a dump truck and slipped. One of the wheels ran over his ankle. They had to take an X-ray, but there was no problem. Then, I myself stupidly stepped on a 16-

penny nail on a piece of 2x4 that was covered with snow. All I needed was a tetanus shot.

Although, on another job, the camp cook was carrying a plate of beans in the mess hall and slipped. He hit his head on the corner of a table and died. Old John Baker's other adage was, "Even the mules know when it's a force-account [cost-plus] job."

Years later, on another dam, we had instructed a foreman to dig an exploratory trench along the upstream toward the dam. This was part of our contract. He completed his excavation, and after the consulting engineers had inspected the material and photos were taken, we refilled the trench with the excavated material and compacted it into the 6" layers.

The old Irish foreman took me aside. "I did not know this was a force-account job."

I replied that it was not a force-account, and he better not operate it as if it were.

He looked at me suspiciously and walked away, muttering it must be a force-account. John Baker worked with a famous dam engineer named Adolph Ackerman, and he spoke about this man after as if he were a deity.

## Sanford Mitchell

Sanford Mitchell was another resident engineer whom I had met in 1938 and again in 1946. Mitch was in his middle sixties and looked older. Mitch was a wise and honest engineer. Fay, Spofford & Thorndike (FS&T), the engineers of the Fall River airport, used him as a fall guy when the job did not turn out the way it was supposed to.

The site of the airport was a high-level swamp. I found out afterward that it had been rejected as an airport site by the Corps of Engineers on the grounds that the top few feet of existing material would have to be removed and wasted. If FS&T had dug one test pit, that was all I knew.

They'd designed the airport as if only three to six inches of stripping would be necessary and the rest of the material could be all used for fill. Apparently, they had only cut enough brush to get a center line through for the survey. They must have used a GS map for the rest of the contour information. Then, they advertised for bids, and the Boston firm of the Dooley Brothers was awarded the contract on the unit-price basis.

Shortly after the contractor started the work, he became concerned about the job and hired me as a consultant on a daily basis. The job soon got so confused, I had to devote about five days a week to it, because the engineers had not properly estimated the quantities.

We cleared the site and found rock, obvious in several places. FS&T had not allowed for the borrow material that would have to be hauled in to replace the unsuitable material; this increased the quantities of fill material required astronomically. The actual quantities required of rock excavation increased so that the cost of the project reached beyond the estimated cost.

The job was half graded, and there was little material left in the excavation to bring the airport up to grade. Faced with a large increase in borrow costs and rock costs, the engineers performed several acts. They issued change orders to eliminate the topsoil, which had to be spread over the seeding area, and they eliminated the rock excavation by altering the approach angles required by the Civil Aviation Administration regulations. They eliminated all the asphalt primer and asphalt paving.

But they continued to demand that the contractor grade the gravel base to the tolerances required for placing blacktop paving. This fine grading is normally charged to the paving price, because between frost and wind, the grading must be done just before the placing of the pavement. They also demanded that the grading of the proposed grassed areas be fine-graded for the placing of the loam and topsoil. These changes left the contractor in a no-win situation.

In order to eliminate the borrow required to bring the airfield up to grade, the engineer proposed to issue a change order to lower the grade of the whole area by several inches, which would require the contractor to skim off the whole area of cut and add to the fill a few inches of the fill areas. This would be done at cost to the contractor, but no new prices would be paid. The contractor refused to sign this change order on my advice.

The consulting engineer finally ordered them off the job and hired another contractor to complete the job. The contractor sued. This suit was complicated by the ignorance of the lawyer, who had no experience in construction. It took me several years to get him to understand the unit price conditions of the contract.

He told me, "Jack, Dooley bid so much. He's already been paid more than that now. How come you want him to get more?"

This finally got straightened out, and Maurice prepared a sort of a case. I testified once a year for five years. Each time I came to court, our lawyer was trying to set me up as an expert. I had books, drawings, etc.

However, the judge happened to be a friend of my father's and knew me. He would say, "I know that Mr. Scheuren is qualified as an engineer, but this is a legal matter, and we do not need his testimony."

Anyhow, after five years or so, Dooley won the case. Because of his lawyer's stupidity, they only paid him the exact figure and didn't pay him any interest. Though he got his money back, he lost enough on the delayed interest so that his equipment rental was free.

My friend, Sanford Mitchell, was removed from this job and put in a menial position in the office of FS&T. They perhaps worried about totally firing him. Later, FS&T, because of their failure on the infamous North African airfield, were to become known as Fumble, Stumble, and Trip. This was noted in the records of the investigation of the fouled-up projects in the North African airbases.

Mitch was a great man–only unfortunate in the choice of the firm he worked for. The change order procedure was an issue many times to cover up. I remember a defense project in the early days of WWII, when the Army and the politicians, just before the Truman committee started work, set up a project with two prime contractors with overlapping work.

The contracting officer at the time, Lt. Col. Joslyn, was smart to see that the job was heading to an impasse and wrote a change order that converted the job, which was eighty percent done and behind schedule, to a hundred percent, so that said 800,000 cubic yards was done and a million yards required. He just changed the figures so that it would be 800,000 yards, 100%. The job showed 100% complete several months ahead of schedule.

Of course, later on, a new contract was awarded to complete the job. But the lieutenant colonel got a medal and was promoted to full colonel.

## Tony Cucolo

Tony was an Italian immigrant who had come to the U.S. and had been successful in concrete paving, plants that mix concrete, and later cross-country electrical distribution systems. He had very strict ideas about conduct—very straitlaced.

We had a small office together at one time, and we used to share lunch at the lunch room. One day, I brought in some seedless grapes.

Tony ate one then said, "Jack, this kind of grape is out of season. How much you pay?"

I think I said sixty-five cents, about twice the usual price.

"Sixty-five cents! That's too much. I won't eat them."

I said, "Tony, at dinner last night, we each had two martinis, and each one cost $1.30…"

"Thatsa different." And he still refused to eat it.

He invited me once to a big clambake of the Electrical Workers Union-High Tension Lines. I asked him how he got into that business.

He said he was subcontracting the clearing of right-of-ways and building the concrete foundations for the towers. The job came up, and he bid it; put up the towers and strung the wire. He said that he didn't know anything about it, but he tried, and he learned but lost a lot of money.

But on the next job, he knew what to do, and now this kind of work is part of his program. He used to say the most important things in construction are price and cost. Never let this information out. He always remembered a job he lost because one of his competitors had a room next to his company in the hotel and overheard enough of his prices to undercut him on the job.

I heard from Tony around the time I went in the Army. He, at forty-six or forty-seven, entered the Army in 1943 as a captain, probably to work in Oak Ridge on the Manhattan Project under construction. He had two sons, Belmonte (Billy) and Tony, both at West Point. Billy had just graduated.

They sent Tony to Belvoir, to the same type of course I had gone through, and he was not well; too old and too fat. But he stood it for a while. Then, he called General Sam Marshall. The general's son and Tony were roommates at West Point.

As Tony tells the story, he said, "General, this is Captain Cuculo."

And Marshall said, "Why great, Billy. You made captain so quickly."

Tony said, "This is a no Billy. This is a Tony. They're killing poor Tony at Belvoir."

Needless to say, Tony was transferred forthwith to boss a bunch of Italian POWs at Tennessee.

He was a deep-thinking man with sterling character. He could speak English without an accent, but if he wanted to make a point, he would fall on the "Poor Tony" routine and get on the other side of the question.

When we were making a deal, he used to say, "Jack, today you wanna be hard or soft or easy?"

What we would do then, if I said hard, Tony would appear to agree with our opposite numbers and then say to me, "What do you think, Jack?"

And I would say, "No." And then we would be off on another track.

## Bert Stuart

Bert Stuart, my uncle and an excellent construction man, made money in Florida in the boom in the 1920s. He quit contracting and stayed in real estate development. He would read the *Engineering News Record,* see some bridge job, and say, "They'll lose their shirts."

I remember in reply once, when he was second bidder on a large bridge job by $110,000 on a $30M job. I said, "Good. You must have that much, and you bid to take that kind of loss with much in your overhead and contingency."

He said, "Of course I have. But we're different."

Even after Bert left active construction, he kept up on the business and visited every job I worked on in the U.S. Construction stiff at heart to the very end.

Then there was G.W. Maxon, Sr., a great figure in the construction business. I was in the Army in Italy with his son, G.W., Jr. I only met the Senior man at a few functions at Lake Moles and the AG meetings.

His son, G.W., Jr., and I corresponded with each other for over forty years. His expression, when he was still in the business, was "checking." Checking meant a great deal, and of course was what a manager of a construction firm must continually be doing.

And other people said, Keith Wason, a Peter Kiewit officer, "Who's going to see to it."

Peter Kiewit said, "I am pleased but not satisfied."

J. Warren Shoemaker, the President of Warren Brothers Road Co., said, "Nothing is done by itself."

Andy Paton said, "We have to not only be right. We have to look right."

And, of course, Louis Perini, "Get paid for what you do."

## Bernt Balchen

Then, there was Bernt Balchen: "Ve do it, viggle the vings."

I knew who Balchen was, his flights with Byrd and so forth. He worked in Greenland in WWII, but I never met him until Thule in 1952, and we became good friends. I went with him on a couple of reconnaissance surveys and a couple of rescues.

I had met him at Ft. Pepperell when he was about to be retired. When we returned on the *Westwind* ice breaker, after a two-month trip, I called on Bert at this office at the base in Pepperell.

He had a large manila envelope. After a brief discussion of our *Westwind* trip, Bert showed me the envelope. "This is all I have left from my stay in the Air Force. At Retreat Ceremony today, I shall be retired." This was 1956.

I went out to the ceremony. I was standing next to a young Air Force captain who asked me, "Who is that fat old man?"

I said, "That's Colonel Balchen. He commanded in Greenland in WWII. He was the first pilot to fly over both poles, and he flew Byrd over the South Pole and also to France years ago."

He looked at me and said, "Byrd? *So*?"

I saw Bernt at a couple of Explorers Club dinners. The last time was when I had come up from Trinidad in the late sixties to "Aviation Night": a $100-plus dinner, planned by the Explorers Club. At the pre-prandial drinks, I met Bernt.

He said, "Hello, Yon," (as he always called me) and introduced me to his wife.

I had Andy Paton as my guest. He knew Bernt from Thule but had not worked with him as I had. At the dinner, Eddie Rickenbacker, Roscoe Turner, and Balchen were all there.

Capt. Eddie slept at the head table all through the talks. Then, Andy asked me to introduce him to Rickenbacker. Captain Eddie was rooming with Roscoe Turner.

At the meeting, Andy and I and Rickenbacker turned and said, "Let's go on the town."

Colonel Turner shrugged and said, "Come on, Eddie, it's time to go to bed." Eddie agreed, and that was the last I saw of him and Roscoe. It was also the last time I saw Bernt.

Roscoe Turner, the legendary "Captain Jack" of the comic strips, was a great character, too. He used to fly sometimes with a pet lion cub around his neck. Roscoe died of cancer not long after that meeting in New York City.

## Maurice Ewing

Then there was Maurice Ewing, the great physicist and ocean scientist. He came to Trinidad to a Geophysical meeting in the late sixties. I showed him around our projects.

He was interested in the stabilization ponds, in our use of otherwise valueless land. When we talked about the core drilling we had done there, he asked for the exact elevation where we had hit selenite crystals.

He told me with shining eyes of a new, undersea sampler he had developed. Then, he spoke of when he had conned the U.S. government in the mid-thirties to let him have a submarine to do some tracing in the Gulf Stream. The submarine was moored in the harbor of Port of Spain. At the same time, President Roosevelt was on the cruiser, *Augusta* at anchor there.

FDR sent a message (after he heard Ewing was there) and invited Ewing and the skipper of the sub for lunch on the *Augusta*. The sub skipper radioed back that their schedule called for them to leave at 1200 that day, and FDR regretted. Ewing was furious. His schedule had been drawn up by one of his kid clerks, and it wasn't a hard and fast one. He never heard what happened to this skipper in later years, but we both figured he never would have made much of a career in the Navy.

At one time in the early fifties, Ewing published a paper in the *Saturday Evening Post* on the coming Ice Age, based on some sediment deposit analyses. Ewing said it was one of the lesser things he had ever published, but it sure got a great press.

When I was on the North River New York job in 1972, I met Ewing at a buffet at the Explorers Club. He had just returned from Prudhoe Bay, for which my firm had designed the oil camp, and was very impressed with the installation.

I remember some character getting into a conversation with Ewing. It turned out he had some sort of a commercial deal and wanted to talk business. Ewing dismissed him with harsh words. He was interested in science. Anything else was chit-chat.

## Martin Lang

All his life, Martin Lang had been in the environment business, wastewater, solid waste, water parks, etc. He wrote once that a handball court in Queens would be as much of a benefit to the environment as a mountain trail in the wilderness. The total environment, as Dubos also said, should benefit all.

To me, the outstanding characteristic of this great engineer was earnestness. Whether the project was large or small, he was interested and dedicated.

I have seen him in action in New York, Washington, Trinidad, and Jamaica. He would charge around the screening area of a sewage treatment plant in Jamaica, New York and ride a garbage truck in Jamaica, West Indies with the same gusto and the same earnestness.

## Father Daniel Linehan, S.J.

Here is a man, a priest, an athlete, a scientist, an explorer, and a friend. We lived in the same tent in Antarctica. He said Mass in my home in Trinidad. We climbed the Northern Range in Trinidad together. We consulted on the rock conditions in Antarctica and in Trinidad. We celebrate our birthday together every five years. On the last one, I was seventy-five and Dan was eighty.

Father Dan was a faithful trail companion in Antarctica. He was on his knees plenty of times, with his bare hands and his developer. We'd gotten one to print his seismic pictures. He was the only one of the group of scientists and engineers I led to Marble Point who completed his work and his reports on time. That is really something for a scientist, as they have a hard time leaving off.

He sounded the top of the ice cap at the South Pole to the earth below, 10,000 feet of snow and ice, from the depth of the rock below the earth color. A down-to-earth operator. He foresaw that wine would be in short supply at our site, and he ordered a bit extra of the sacramental wine, which he distributed to us, to help the mood.

In Trinidad, we gave a dinner at the country club and had Dan give a lecture with slides about this seismic work. Some of the local doctors were astonished to have a priest explain science. One of the local reporters interviewed him and asked about science and religion.

Dan gave a good interview and explained that there were no differences between science and religion—only between some scientists and religion.

## Dr. Robert L. Nichols

I was several years in the Arctic and Antarctic with him. He, like Dan, is a great trail companion and was one of our party at Marble Point.

Bob spent a year in the Antarctic in 1946 with Finn Ronne. The ship they had was an old Navy deep-sea tug. It was named *Port of Beaumont* for that cruise. Coincidentally, we used the same ship in 1957, after it had been bought by a Canadian outfit and used as a sealer and was rented to the USAF at that time.

The *Arctic Sealer* was, of course, with us for three months in Greenland. Bob was always a real tiger in the field. He tried to carry big loads like a Sherpa. And when he was driving dogs, he used to pull along with them. Another giant of the Earth.

## Millard Mark Klinefelter

You meet many people as you go on this Earth. Some are ordinary, some are interesting, some are great. But only once in a while do you meet a man who sparks against your own character like flint against steel.

In 1943, at Fort Belvoir, Mark Klinefelter and I were taking a Basic Officer Leader Course. This was a course designed for civilians who had been given direct commissions for some specialty. For sixty days, we would run through the basic training of enlisted men and the ninety-day course for officer candidates and so forth. So, we were kept on the run.

Mark and I are about the same size, configuration, etc. We both have similar complexions and eye color, and we were both mistaken for each other several times. We were both construction stiffs. Both had started to work practically as children. On the way, I picked up a degree from MIT. Mark had finished school to the tenth grade, but he had filled nights for years with correspondence courses.

The physical torture of the training was no problem for either of us. During the course, when volunteers were called to jump into the barbed wire—for two to jump on your backs as stepping stones to clear the wire—we both were there. If someone had to crouch in a tank trap while tanks went over the top, we both were there. Obstacle courses, infiltration courses, whatever we were called upon—no problem.

One day, the young Tac (for tactical; like drill instructor in Marines) officer got annoyed at something and ordered all of us to run. He said, if anyone got tired, he could drop out.

Mark and I kept right on the heels of this young squirt. We were ready to fall on our faces with fatigue, but we kept going until the Tac officer finally called it off. All the other students were, by this time, sitting down under the trees.

Then, we went to Fort Leonard Wood. It was very boring there. I was teaching the recruits the techniques of demolition, explosive mines, booby traps, etc. Mark was teaching rigging and all that. He had been a deep-sea diver, and I had little experience at that. I had been under air, but it was as a sand hog (in tunnels, caissons) under compressed air.

When we saw the circular asking for candidates for diving underwater in Florida demolition school at Fort Florida, we eagerly volunteered. It was better to be at Fort Leonard Wood without the snow and rough out there. I ended up as a student at Cooks and Bakers School, and Mark as a radio announcer on the rifle range.

We went to North Africa as part of a cadre that was supposed to officer a regiment of cavalry that was remade into engineers. By the time we got there, it had been decided they could use the same cavalry officers as engineers, and we were all dropped into the replacement depot. Several of us, including Mark, were ordered to take command of firefighter platoons: a small fire department, one lieutenant, and sixteen enlisted men, some fire trucks, etc.

At that time, I had a small operation scheduled for a barbed wire accident in the U.S. I missed the first draft of the fire department people. Mark went.

Then, I met Jim Lambert, an old lieutenant colonel in the Corps of Engineers, who had been a friend of mine at Perini years before. He got me transferred into his outfit, and Mark's platoon was part of Jim's command.

One of Mark's corporals was drunk one day and couldn't drive his truck. Mark didn't wish to court martial the man, so he restricted him to his quarters. The man broke the restriction and went AWOL.

I was on the court. There was no defense. The soldier pleaded drunkenness as an excuse but still got six months hard labor. The Disciplinary Training Center (DTC) was operated by a really tough MP Major, a former prison warden and sheriff. His camp had a practically zero-percent return rate. Once out of that place, they straightened up and had no desire to return to his custody. In that DTC, all day, they drill with full field packs and shine mess kits to make it shine like a mirror; or break rocks into little ones. There was no physical abuse or torture, just keeping them busy.

Then, Mark's outfit was ordered to move. When you have a man in DTC custody, if you move, you have to take him with you—in custody—and then turn him over to the DTC at the next station. Or you have the commanding officer in the area remit his sentence and put him back on duty. I went to Col. Lambert with the papers and got him remitted.

I went down with Mark to get the man. He had had a big moustache, but when we picked him up, his moustache and beard had been shaved. He was in good shape, but he thanked us profusely for getting him out. When the ammunition ship blew up in Bari, Italy, he, as well as Mark, got Silver Stars for carrying out the wounded and fighting the fire, with 50-caliber rounds blowing up all around them.

Then, Mark and his outfit went to southern France. He was operating the firefighting platoon at the docks in Marseilles. I came over and went to his area.

He didn't even look up and seemed to know who I was without looking. He said, "You know how I could get out of this job? Should I write to Marshall Eisenhower?"

I don't know to whom he wrote, but he finally got transferred to a construction battalion in Okinawa. He got a spinal injury while diving for a sunken ship, so he left the Army with a medical discharge.

He next went to Venezuela to be a steel superintendent of Creole Petroleum. While he was there, his Okinawa injury cropped up as an abscess in his vertebrae. He went back to San Francisco on active duty in the Army hospital there. When he was healthy again, he went to Korea as a major, to build airfields. Then, he got transferred to Austria and the Engineering Command.

Next step, he was made area engineer in Norfolk, Virginia, which was in the NE Engineer district, where Mort Solomon was in charge of the Greenland work. I asked for him to be sent as an M&E civilian to do our work in BW-8. This caused a flap, as the civil servants in the district did not wish to have him appointed as Chief Engineer at BW-8, because he had no degree, only completing tenth grade in high school. Even Andy Paton fought me on this.

I went to Col. Solomon. Mort looked at his papers and said, "Perhaps he should be working for the contractor."

I said, "So should I, but I need this man, and you will never be sorry."

Mort got him relieved and out of the Army, and he went to Greenland for M&E. He did some splendid work. Later, even Andy said Klinefelter was the "best thing that happened to us this year."

Construction gained on him, and after he went and had done a splendid job for several years, he told me he wanted to leave. He wanted to recommend Russ Holt (later to become the Executive Vice President of M&E and is now in Egypt) be sent back as his deputy for a year. Then, Mark would depart, leaving Russ in charge.

This took some doing, because Russ was head of civil design in Boston and Harry Kinsel did not want to part with him. But I convinced him that Russ's value would be enhanced, which it was.

Mark went to Iran to work for J.A. Jones, a big contractor. He admits his main reason for wanting Russ was that Russ was the only one who came visiting the job from the main office who didn't have a camera swung around his neck like a tourist, and Mark thought that meant he was a serious worker.

I only knew Mark for a little less than thirty years, but he was always an example of a fighter, a dedicated construction stiff, and dedicated to getting the job done.

## Red Collier

I want to talk about George Herman "Red" Collier. Up to now, I have described friends and people whom I admire, but it is about time I depicted a man who is certainly not my friend and not one I admired. He was a character—I knew him well.

Red was a football star at college and played a bit of pro ball. He had worked a little in pipelines and then become superintendent of schools, I think in Seminole, Oklahoma or maybe Texas. While on this job, he was caught with his hand in the till, and afterward, when prosecuted, he was ousted from the academic profession.

He was literate and had started a pipeline magazine with Hobbie Dunn, pipe-stringing contractor. Stringing is the portion of a pipeline that unloads 40-to-80-foot lengths of pipe to flag cars that load them on pole trailers that carry the same load as telephone poles, and they distributed the pipe sections along the right of way. Besides his journalistic efforts, he went on running pipeline jobs as a foreman or superintendent, which they call *spread boss* on the pipelines.

When I first met him in 1941, he was pipe foreman for Foreman Company, who had a 26-inch pipeline in Kansas-Nebraska. He was an excellent pipe foreman, but when we transferred the spread boss to another new job, I made Red spread boss. He was about 6′4″ and 240 pounds. Like most big pipeliners, he wore a big Stetson hat, big yellow Texas cowboy boots, and big yellow leather gloves. He drove a big Buick, and he hauled his wife and son, George Herman Collier, Jr. (a red-headed boy about ten years old at the time) around in a house trailer from job to job.

He ran the job very well. He was a heavy drinker (not on the job) and a ferocious skirt chaser. Pipeliners are not any worse or any better than any other itinerant working stiff. But Red was more obvious than some. I remember having breakfast with him at a restaurant in Concordia, Kansas. The owner/operator was a fine-looking lady, whose husband was off in Topeka and was a state representative.

There were only three of us in the place at 5:45 in the morning. Red chatted with the lady and then said, "I believe I'll go back and make sure those eggs are scrambled right."

So, the two of them went into the kitchen, and after a while, they came back with the eggs. But both Red and the lady were a little flustered. When we left, I asked Red what took him so long?

He said, "Well, we talked about love, and she flattened out, so I cut her."

I jokingly said, "How would you like it if your wife was carrying on like that?"

"I'd kill the bitch," he said.

Later on, I had a job in Chicago. Red was driving his big Buick out in Cicero when a Cadillac cut him off the highway. He chased the Cadillac, pulled it over to the side, and walked up to the driver

He said, "You son of a bitch, I'm gonna make you s**k me off." And he was trying to drag the little man out of the car when he had a gun stuck in his ribs.

The man had a girl with him, and he told Red, "I'm gonna blow your guts out. One more dead rube will mean nothing to me."

The woman said, "Ah, leave him alone, Louie. Don't bother the punk."

The gangster told him, "Get out, get back in your goddamn car, and screw. If I ever hear of you around this town again, I'll blow your head off."

Red had to change his underwear after that, and he perhaps was more careful about challenging some little man who might have an equalizer with him.

Then, we had a big pipeline job Red was the spread boss on. When the job was done, there was no big job to go to, so we let Red stay there with a small crew of welders, etc., to clean up the odds and ends in drips and blowout valves on the natural gas line. He tried to string out the job. I went and found them all just lying around and dogging it, so I told Red to take his company car into Kansas City and get paid off.

He just said peacefully, "Okay, Johnny," and left.

Later, when I was in the Kansas City office one night, estimating a prospective river job in Nebraska, Red called from Borger, Texas, which is the center of oil and gas activity.

He was lousy drunk and was abusive on the phone about my firing him. The only good thing he could say about me was that he called me a "Boston Baked Bean." He told me he would be in Kansas City the next day and wanted to know where I would be, so he could whip my ass.

I went through the usual pipeliner dander and told him, if he whipped me, he would be getting no maidenhead, that I had been whipped by better men than he. I knew I couldn't beat him in a fair fight, but I hoped I knew enough dirty tricks not to give him a fair fight. I told him I would be in our warehouse the next day, and if I had to leave, I would tell Chief West where I would be.

The next day, I went to the warehouse and told the Chief the story. The Chief was about Red's size and a famous rough-and-tumble fighter. He said, "Do you want me to whip him for you, boss?"

"No, I'll do the best I can."

Anyway, Red never showed up.

Many months later, I had to replace a pipe foreman on a job in Missouri. The spread boss had gone phony, and I had fired him and put the pipe foreman in charge. This pipe foreman, Jack Henry of Bonham, Texas, was a really fine gentleman (rare among pipeliners). Jack finished the job but died of TB not long after.

Anyway, I called Kansas City and asked C.S. Foreman if he could find me a new pipe foreman. He laughed. "I have a good pipe foreman in my office right now, but I don't know, he's not sure whether you want him or not. It's Red Collier."

I told him I would not want him as spread boss, but I would love to have him as a pipe foreman. In a few days, Red and family and trailer arrived. Red walked across the lot, and I asked him, "Red, do you come in peace or do you come in war?"

"Ah, I was just drunk that night. I didn't mean any of that." He set in and did a good job.

When the War Emergency Pipeline in Ohio was underway, I met Red in Lancaster, Ohio. I was just leaving to go into the Army.

Red said, "You must be crazy. Hell, I'm making so much money these days, I hope the war lasts forever. I'm in a good thing."

I never saw him again. But some years later, when I was going the jobs in the Berkshire area, I met one of Hobbie Dunn's stringing foremen (he was stringing a line for Shell). He asked if I'd heard what happened to Red Collier.

"He's dead. His bronc (*bronc* means son in pipeline lingo) shot him."

The boy had been about fifteen or sixteen by that time. His father, Red, came home drunk one night and started the normal beating of his wife. Herman, Jr. took a shotgun and blew the old man's head off. I don't believe he was ever charged for it.

So, that's the story of Red Collier.

♦ ♦ ♦

# CHAPTER 2

## Early Jobs

Tom Murray was a greenskeeper at the local Swampscott County Club when I was a little boy of eight or nine. He hired me to dig dandelions. We used a paring knife for dandelions and a fork to dig witchgrass out of the greens on the golf club. It paid fifteen cents per hour, which was the same as caddying, but it was steadier work, although you got no tips.

Tom was an Irishman. I would have said an old Irishman then, but I guess now he probably must have been in his late thirties when I was first hired for summer work. When you dug weeds, you had to work on your hands and knees. If you sat down, Tom would tell you to get up on your knees. His exhortations were to "bend your back, boy."

I worked there several years during my schooling, even including several of my MIT vacations. The country club was engaged in building an additional twenty-seven holes for golf and several tennis courts, and old Tom was in charge of construction.

I grew along with the program, mowing grass, raking sand traps, clearing brush, digging ditches, laying pipe for water and drainage, and helping the masons, and I rose to be driving a team of horses at about thirteen years old. Construction was still using horses and mules then. Ford Model T dump trucks and Fordson tractors did some of the work.

I drove an old horse named Chubby, a gelding on a tipcart, and sometimes Chubby and an old mare called Tessie on a stone drag/bottom dump wagon. The tipcart work was better than the digging of ditches.

Boulders and ledge rock were drilled and blasted, and the fractured rock was loaded by hand into the dump cart. Teamsters had to back the cart up to the rock piles and join the laborers in filling the cart with rock, then drive the horse and cart to the place the rock was to be dumped–a tennis court foundation or a base for the green or for slope paving or riprap on the edge of the water hazards. There was a hook to be released after you got your horse and cart backed into the right place. Then, you released the hook and by brute force lifted the end of the box so the broken stone slid into place.

Old Chubby had a much longer time in this practice than I, and there was no need to use a whip. (I didn't even have one.) He would walk along then turn back and stop, so I could lift the end of the tipcart and rehook the box, then walk back to the digging area. Chubby was an old horse when I met him, and after several years, he was failing in health, so Tom decided to put him out to pasture. The next step would be the last for Chubby–the glue factory, where he would be executed and made into dog food, his bones into glue, and his hide into leather.

The last day he worked, I led him to his stall and got him fed and cleaned up. Being a teamster caused one to work longer hours to take care of the horse, but of course, you didn't have to do so much digging and shoveling. I still remember how, on the next morning, I had left Chubby in the stall, but he got himself out and went over and backed himself into the shaft with a tipcart waiting to be harnessed. Poor Chubby.

At that time, I was getting a man's pay for a man's work. We all got fifty cents an hour for five ten-hour days and one-half day on Saturday, $27.50 per week. This pay never increased over the next five or six years I worked on this job, but it helped me go to MIT with my summer earnings.

I was fifteen years old at this time and working with a rock crew. We had a blaster who used to get drunk sometimes. One day, the blaster did not show up for work. Old Tom got him out of the drunk tank at the jail later.

However, the rock had to be blasted. The rock was drilled by double-jack method, by hand. One man held the drill against the rock, and the other hit the steel of the double-headed hammer or jack. There was also a long-handled spoon to clean the cuttings out of the drill hole. This was just about the time, of course, that horses were being replaced by trucks and tractors. Companies like Ingersoll-Rand were developing modern air-operated equipment. These air drills were called "jack hammers," and it is a trademark of the company to this day.

I had been helping the blaster, so Tom asked me if I could break the rock with a mud cap. I had seen the blaster do this and, being a cocky kid, said sure. I got the powder. Black powder had been used for so long that blasters called any blasting agent "powder" and still do.

As I had seen the blaster do, I cut the cartridges of dynamite in half and packed the powder on top of the boulder, taking care to make a slight indentation in the center of the mass of the relatively plastic dynamite. (It was many years later, as an engineering soldier, that I was to use "shaped charges," which became part of the Army's demolition equipment.)

I covered the explosive carefully with mud, as I had seen the old blaster do, but the cover looked a little light and flimsy, so I added a few chunks of broken rock to hold it down. Of course, after I lit the fuse to detonate the blasting cap I had placed in the charge, the boulder disintegrated satisfactorily, but a piece of the rock cover went through the roof of a Model T sedan a few hundred yards away. This cost the country club a few dollars for repairs. Since then, I have handled or directed the handling of many types of explosives, military and civilian, in many parts of the world without any untoward incidents. But in this case, I had to learn the hard way.

The work with Tom was rewarding. The men I worked with were pleasant to me and constantly drew my attention to their plight, that of the unskilled and relatively unskilled worker. They used to tell me, "Get an education, boy. Don't be like us, who only can find work with the Irish anchor (pick) or the Italian banjo (shovel)."

As I look back, the approximately ten years I spent on that work really made up a large percentage of my working life, almost sixteen percent. In later years, during the Depression, when the country club abandoned the extension projects, old Tom came to see me. Only half-heartedly, he reminded me that he could still make them "bend their backs."

When I was at MIT, I worked at several jobs. I entered as a student of civil engineering. Most courses at MIT were much the same for all freshmen. However, I changed courses at the end of my freshman year, because the civil engineering course required a summer course on surveying in Maine. This would cost money and keep me from my needed summer earnings.

Besides summer construction jobs, I worked part-time after school in the Bursar's Office as a clerk, at the Foundry, and as a janitor's helper. All these jobs paid fifty cents per hour. The janitor's helper meant cleaning toilets, fixing light fixtures, and mopping floors.

The janitor, Tucker, was a czar in Building I. He got less money per hour than his student helper, $19 per week at forty hours a week. Tucker was a stickler for quality work in his chosen profession. He would judge a man by the way he handled a mop. I can always remember his telling me, "Mr. Scheuren, (he always called his student employees *Mister,* and we always called him Tucker), you have taken to this work very well but…" He sighed. "Mr. Smith will never understand the mop." There can be pride in working at college, even at mop handling.

Another job I had during my college years was working at the Roseland Ballroom in Boston's Back Bay. I had to wear a tuxedo, dinner jacket, black tie, or whatever and be the bouncer; security man, I suppose it would be called today. This job paid $1 an hour, four hours an evening.

This was no dime-a-dance place, the boys paid eighty cents each and the girls, forty cents for admission. It was not the same type of group you would find at fraternity dances, tea dances, or balls at the big hotels. This group was, for the most part, boys and girls who enjoyed dancing to Jimmy McHugh's Orchestra. (Jimmy later went on to bigger and better things.) They were what the Alger books would call respectable, poor but honest people. The only trouble we had was from "slumming" college boys, who had improper feelings about the virtue of these girls. (Remember that the girls had paid their own way in.)

I remember one incident when a drunken student from Dartmouth turned up and asked a girl to dance. Of course, his accent betrayed him, and besides, he was drunk. She looked him up and down from head to toe and said "no thanks" and turned away.

The student grasped her arm. However, I collared him and got him to the door, got his coat check, and pushed him out into the street. All the time, he was telling me he was Billy Alton who was Dartmouth's 160-pound boxing champion. As I was on the MIT boxing team and boxed the Dartmouth team that afternoon, I knew he was not Alton, but I said nothing except "get out."

As it happened, there was a big Boston policeman passing by, and the student bumped into him. The patrolman straightened him up and said, "Come on, boy. Take it easy. Just move along."

The boy looked at the policeman, whose badge happened to be 1620, equivalent to a historic date, and started arguing and wrestling with the cop, saying, "I got your number 1620, and I'll have you busted," or words to that effect.

The policeman looked at me apologetically and said to the boy, "I don't want to get you in any more trouble," but after the student reiterated his threats, the policeman called the wagon, and the stupid young man spent the rest of the night in the drunk tank. But usually, the work was peaceful and pleasant.

Then, I decided, instead of graduating with my class, I would get a leave of absence and work a whole year, so I could earn money enough to go to school as a gentleman and not be working nights and Sundays. I went to see Colonel Charles R. Gow, who at that time was postmaster of Boston and a professor at MIT. He was a professor of a new course in humanics, what is now called humanities and later taught by Professor Magoon. It was supposed to humanize engineers and later became part of the courses in Industrial Management mentioned in my paper, "Engineer as Entrepreneur."

Colonel Gow was also President of Gow Co., which was a foundation company usually engaged in the use of caissons and foundations. He was also vice president of Raymond Concrete Pile Co., which had been taken over by the Gow Co. as its subsidiary. He offered me a position as a rodman on the caisson foundation of a new telephone building his company was doing in Boston. The superintendent of this job was Michael J. Moran, then about fifty-two, a fine gentleman and an experienced engineer and construction superintendent. He was then a bachelor and had traveled all over the world.

The Gow system called for hollow steel cylinders, 6-feet long, in varying diameters. On this job, the depth to rock was thirty-five to forty feet. The shell would be set on top of the ground, and caisson workers excavated inside the cylinder, so it would be sunk. After one was about six feet down, another one, an inch or two less in diameter, would be slipped inside and excavated, and so forth. The excavation would be continued into the rock and test drilling made, to make sure the rock was continuous.

Then, the hole would be fitted with concrete and the cylinders withdrawn as concrete rose. The cylinders were, of course, reused. It was my job to work with the engineer and check the quality of the concrete before placing and check the competence of the rock at the bottom.

Men, including myself, were let down in the hole by means of a hand winch and rope tied around one thigh. Excavated material was brought up by the same hand winch. The adjacent buildings had to be braced and underpinned. Part of the job was to check the building for any movement, either vertical or horizontal. This was a continuing process, and the records were carefully kept.

The company had an excavation near a nightclub a few years back, and this Pickwick Club had a real disaster: many people died when the club fell down. The problem was a weakness in the design of the building itself, and the collapse was attributed to vibrations caused by the dancing and because of a great number of people on the floor combined to cause the failure.

But of course, the Gow Co. wasn't taking any chances. The caisson workers were all young Italians. They came to work each day in dark suits, white shirts, and dark neckties, carrying canvas bags for their work clothes. The caisson superintendent was Nick Fiore, a fine Italian gentleman who ruled his gang. He was called the Kaiser. My boss was an engineer from Kentucky. Very meticulous about his work, but did not shave enough to suit Nick the Kaiser.

Nick used to rub the engineer for not shaving, saying, "You shame-a the job."

"You don't shave everyday yourself," retorted the Kentuckian.

"That's-a different for me," said Nick. Anyhow, the Kentucky boy did not last long.

We had points established on the sidewalks of the two sides of the excavation and on the walls of the building on the other two sides, in order to locate the caissons. There were sheetpiles to be driven, and the pile driver foreman chose to set a steamboiler on top of one of the sidewalk points.

He and the engineer got into an argument which ended in a fistfight, and the engineer not only lost the fight, but he also lost his job. The manager of the construction project believed that the engineer does not make any money for the job but can lose it, and therefore the pile drivers had already spent the money to set the boiler and the engineer should have worked around it.

Mr. Moran called me and asked if I could take over the engineer's job. He showed me the field books. I had just been holding a level rod on a ladder and of course had never seen the books.

After looking at them, I was nineteen then, I told Mr. Moran I did not believe I could read the rod readings through a transit across a large area with steam and vibration and get down a thousandth of a foot.

He said, "Of course you can't." And neither could he. If we had any trouble and had to go to court, it would be bad to show this book. "I don't expect these books to be kept like this. If you find 1/8" out, check it again. If you find ¼" out, come and tell me right away."

The promotion got me a small raise, but it was to cause me trouble later. All the lines on the grid were laid out on a 90° angle except the one where the pile driver had set his boiler. That line was 89°, 59 minutes. When I had to lay out the caisson on that line and went around the boiler by running 90° lines, I had not carefully checked the drawings. Anyway, just before we finished placing concrete on that caisson, I found the centerline was 1-1/2" off.

Of course, I immediately went to Mr. Moran. He was in a meeting with his clients (AT&T engineers). No one spoke but Mr. Moran.

"What size had we bottomed out on?" It was one inch larger than the estimated size. He said, "We have plenty of bearing. Can you set the anchor bolt exact?"

"Of course," I said.

As I scampered out, the chief engineer of the telephone company said, "You'll be a lucky man if this is the worst mistake you ever make." He was sure right.

When the job was finished, it was nearly spring. I could have gone with Moran to New York but chose in stay in Boston, where I had an apartment, and then go back to school the next fall. I went looking for a job.

In 1930, there were not many. The *Christian Science Monitor* was building a new publishing facility. I was told to see a man at the construction company. It was a short meeting. I had on, of course, my boots, my riding breeches, and a big felt hat, like all good engineers wore then. This man looked at the small résumé I gave him.

"Are you a scientist?" he asked.

I said, "No, I'm studying to be an engineer." The case was closed. I was ignorant of the meaning of the "scientist' to him.

I knew there was discrimination in Boston. When I was five or six years old, I would sit on the kitchen floor and read the paper, the *Boston Transcript,* back to front. I still read papers that way after a glance at the headlines. As the help-wanted ads used to say "No Irish Need Apply," "Protestant Only." I asked my father, why Protestant only for servants? And he said, because his employer does not want him to leave to go to church on Sunday.

Then I met a man who told me J.H. McNamara Co., a sand and gravel provider that was starting a transit mix concrete business in Allston, needed someone to run his plant. I said I didn't know anything about transit mix plants, although I had been checking the concrete of Boston Sand and Gravel, which was plant-mixed and distributed in a bathtub-type body with agitators.

"That's all right," he said. "Johnny Mac doesn't know anything about it, either."

I went and got the job. It took me a little time to get the mixes right, but we made it, and I made lifelong friends of the old man and his sons, Albert and George.

Three things stand out on that tour of duty. We had an order to supply concrete to a combined office building and nightclub. This was a cash-on-the-line job, and the truck driver presented the ticket and got the cash before he dumped the load.

The nightclub operators then, and perhaps still do, cuff as much as they can, build the places, and then go bankrupt. A new corporation with the same principals takes over, and the creditors get paid twenty-five percent. The nightclub owners get low prices.

This was the same thing, but what happened was the owner wanted certain concrete for some finish inside the club to be green. Concrete colors were not in stock in those days, so I designed a mix of some yellow ochre and Prussian blue to get the right green. When the truck came into the plant, we filled the mixer with the green mortar.

Then, the truck engine blew, and we had no mixers available for a long time. The only dump truck available was an old Bull Dog Mack truck of George Breen with an ancient raddled body. We towed the mixer up a ramp, backed the old Mack under the chute, and filled the dump truck with green mortar.

From Allston to Park Square, the green mortar dropped over Boston streets, but enough got there to do the job. This was 1930, and the nightclub was the Coconut Grove.

When I went back to MIT that fall, we hired a plant operator named Bill Young to take my place. In 1934, he came to work for me in Arlington to run a concrete plant for Perini. In 1942, as a Coast Guard recruit, he died in the holocaust at Coconut Grove.

In the fall of 1930, I went back to MIT. I took enough money out of the bank to pay my first semester tuition. Shortly after, the bank failed, and depositors got about twenty-five percent, some fifteen years later. I had to go back to work nights, Sundays, and holidays to finish school. I got a small loan from MIT, which I paid back in a couple of years.

One of the jobs I had, number two on my list of standouts during my McNamara tour, was to make tests each week for Mr. Mac on his concrete. I was doing my thesis on transit mixed concrete plant location and used to take the samples over to MIT laboratories for testing. Mac paid me the same fees as he would a regular commercial laboratory.

One snowy day, in the winter of 1930, I was in the plant office in Allston with J.H. and his secretary. Business was down to nothing. The phone rang, and the girl answered.

"Why yes," she said, "we have all kinds of sand."

Old J.H. said, "Pass me that phone." The old man put on his charming tones. "Yes, madam, we have the best washed and graded sand, brick sand, filter sand, and concrete sand." Then, he stopped to listen. He had a very high color. He'd never taken a drink in his life, but he looked like it, and he kept getting redder and redder. Finally, he said, "Madam, I recommend you throw that cat out in the snow," and hung up.

For a while after that, we used to say that we couldn't sell anything—not even enough sand for a cat box.

The next Mac event came when J.H. called me and said a contractor on the paving of the Boston-Worcester Turnpike wanted to demonstrate to the State Highway Department the use of transit mix concrete on the highway. He'd arranged to use JH's plant to mix the concrete that was to be tested at the South Boston Testing Facility of the Highway Department.

I went out and checked all the scales and so forth and personally weighed out the mixtures. And I met Louis R. Perini, the president of Perini & Sons, the contractor. He climbed all over the plant with me and carefully checked all of the mixers we had there. He asked me what I was doing. I told him I would graduate from MIT in June.

He said, "Come and see me then. If you want to work for me, I would give you a job."

I had a successful interview with Joe Maney, later to become associated with Perini on a dam job and with the ownership of the Boston Braves, for Sears Roebuck, sort of a cadet job in South America. Joe wanted me to do some building work and taking off quantities and so forth.

Anyway, I went off the Turnpike job and did not find Louis Perini for almost three weeks. During this time, I traveled with Jack Guiney, who had been an independent truck owner at Mac's and who had a transit mixer mounted on his truck and was hauling for Perini. Ted Guiney, Jack's cousin, a master truck mechanic, and his wife took me in, and I stayed for a time in their house in Ashland. I rode trucks, looked at plants, watched the paving, and watched the pipe work, excavation, and so forth. I met Jim Lambert, the shovel boss and Len Williamson, the general master mechanic and also got to know many operators and drivers.

Then, Perini sent me to Wellesley Hills for a job on the turnpike, which had been held up because of delays in getting the Boston-

Worcester Street Railway removed. In Wellesley, we had several gravel pits on the job, and upon checking, we found that the total was exactly the same as our measurements, but the chief had fixed up the figure so that each plant was different.

One of our jobs in Wellesley was to clean up a real tree garden planted by a real tree expert. These trees were dug up and moved to Boston for the Christian Science Building, hoisted over a high Indiana limestone wall, and planted and guaranteed for two years by a half-mad black Irishman named O'Holloran. He used no equipment except a truck crane and trucks. He used to say, with a rope and a bar, you can do anything.

On cleaning up that job, a job I actually started and left for a season because of the land taking problems. It was the only job I ever did with Perini that I started and completed. I used to start them and then, after the rough work was done and the river diverted or tunnel holed through, I was moved off to start another one.

This time, I moved to a small road and bridge job in Haverhill, New Hampshire. This was really my first job as superintendent by myself. I was twenty-three or twenty-four.

Louie Perini said, "This is the first job for you on your own. All I can tell you is that I want you to get paid for what you do. Keep good records. And if you dig ten yards of rock, I want you to get paid for ten yards of rock. I do not want you to try to get twelve, and be sure you don't settle for eight. And that's it."

On the Haverhill job, I had several foremen. Nicoli for the gravel crusher plant, Rocco for grading, and Fortier on pipe. Nicoli, who was Louie Perini's brother-in-law, was a great builder and operator of sand, gravel, and concrete plants.

We had two basic problems on the job, and one was at the gravel pit. It was an outwash plain of the overlain brook, and the material was streaked with clay, which has to be carefully mixed with sand and gravel to meet the specifications.

The other main problem was Gale Carter, the State Inspector on the highway. He had never done any roads except concrete and this was to be penetration macadam. He really loved to talk. As the job neared the end of the gravel spreading, it was necessary for me to spend all my time keeping him away from the graveling operation by asking for his advice about the pipe laying and of concrete. I listened to his tales and did much of my work during the evenings, so I could keep him away from our gravel blending operation.

As was to happen again and again in my career with Perini, I was transferred to a new job in Arlington, Belmont, on the new Concord Turnpike, and a new man with the company, Jim Tidyman, came up to finish the job. Jim was a real serious engineer and hard worker. I told him about the gravel situation and advised him, for a few weeks, he should keep Carter away and listen to his yakking.

He disagreed violently and said he wasn't going to listen to that old man and that he had work to do. He let Carter see what was going on and ran into trouble. Carter would sit in his car in the pit and reject any load that was not to his ideals.

The general superintendent got hold of me on the job in Arlington, and we fixed things up. I had an ace in the hole, a sand and gravel bar in the stream that would be satisfactory, but a little more expensive to be loaded and hauled. So, that got straightened out.

Then, the broken stone for the paving surface came in. Rocco was one of the best men on penetration macadam in the business. He spread the stone, spotted and timed it, rolled it and spotted a little bit, and got ready to apply the asphalt. Carter had only heard that the secret of this type of paving was rolling, so he told Rocco to roll it again and again.

Rocco told him it was wrong but the rollers kept working until the stones were practically pulverized and powdered and the asphalt when applied lay on top like a blanket—and of course there was no penetration. Perini complained to the management, and Gale was transferred. A new inspector who had experience with penetration was sent in to complete the job.

I set in at Arlington. On this job was a large box culvert, a bridge over the railroad, and a highway bridge grade separation at Pleasant Street. The pavement on the main road was to be concrete, and the access roads and ramps were macadam. We had to excavate large quantities of peat and refill with granular material. We also had a large traffic circle joining at the new highway with Alewife Brook Parkway.

Perini had arranged to haul fill from an old brick yard, clay that was not quite good enough to make brick. This was hard to handle and had to be mixed generously with sand from another pit. My old friend, Mac had an old gravel pit a little farther away that was almost worked out and was now being used as a landfill. But there was enough left around the edges to make this fill and not have to use the clay.

Old JH wanted more than Perini figured he should pay, when we considered this slightly increased length of haul, the difficulty of hauling the clay, and the cost of Mac's material. Louie dispatched me to talk with JH. This fine gentleman treated me as a dear friend, but as I told Louie, there is always a point where friendship ceases, like price, and Louie didn't come to terms. The price of the material had not changed.

We set up a concrete plant that my friend, Bill Young operated. The highway required a large fill along the edge of Spy Pond in Arlington. This fill, which was mostly sand and gravel, displaced some of the bottom, mud, and so forth, so that islands appeared in the pond. In order to remove these islands, we engaged Baystate Dredging of East Boston, which later was taken into Perini Corp. as its Marine Division. Baystate brought in a sectional portable dredge and pumped the mud to some low-lying land not far away.

David J. White was the president of the dredging company and later became Chairman of the Board of the Marine Division of Perini. Dave White was born the same day as my father, and he grew fond of me and I of him over the years. I had known him before, but this was the first time I worked with him.

After the dredging was done, I ran the old steam crane Baystate had used to dismantle a section dredge, then I was to rent a clam shell bucket to do some other work. I went to Dave's yard, saw the bucket on a deck scow, and we made an oral agreement.

The bucket did not show up, and I found one in Perini's yard in Framingham and made do with that. Then, one day, Joe Perini, the treasurer of Perini, called me up and said we had received a bill from White for the rental of the bucket. I think it was less than $200.

So, I called Mr. White, and he said that his bookkeeper told him it was shipped. He said we should talk about it over lunch at the Engineer's Club in Boston.

I went to the club, and Dave bought me a drink and lunch and then said, "Let's get down to business."

I said we had looked at the bucket on the deck scow, but that I had never received it.

He said, "Okay, give me the bill." He took the bill and wrote paid in full on it and probably chewed out his people the next day. He was a great and fine character.

We had in our garage in Arlington many ten-yard trucks, six-wheel trucks. This next winter, we got sixteen inches of snow. The authorities put the blame on the weather forecaster, Mr. Rideout, because he had only predicted half that. By this time, we were shut down and the trucks were not registered.

I went to see Billy Morrissey, who was the Pavement Commissioner in Boston, later to be Commissioner of MDC and have a boulevard named for him. The city had several contractors getting paid for removing snow by the cubic yard, but they were falling behind.

I made a deal with Billy to furnish so many trucks by the hour, if he would guarantee enough hours to pay for the registration and insurance that would otherwise not be required until April 1. I made the same deal with the Town Council in Arlington, to furnish bulldozers and trucks to them, and also in Lexington and Concord.

I had a 1934 V8 Ford with chains, and I was able to cover the countryside. We used to supply each truck driver with a four-ounce flask of whiskey. This would keep the drivers on the trucks. JH used to do that with his plow people, otherwise the driver would go into a barroom to get warm and get drunk. Really, alcohol is not a protection against the cold, but mind over matter.

This was the first job that Perini had not sublet the bridges on. We had an excellent bridge man named Randall and a great carpenter foreman, Joe Hachey. Louie Perini asked if I could one day get along without Randall. They would send him to build the bridges on the Concord end of the turnpike. I told him we could get along without Randall; Hachey and I could finish. So, he went to Concord.

Randall was an amateur magician and a good one. He tried to teach me a few tricks: making cards disappear, how to drive a 20-penny spike with the palm of your hand, and how to hold a pool cue at arm's length with two fingers on one hand. One day, I agreed to pick him up at a garage where he was having some work done on his car. At that time, Perini's superintendents all drove their own cars and were reimbursed.

I was delayed on the way, and when I reached the garage, Randall had been showing the mechanics some card tricks for over half an hour. When we drove away, he told me, "Boy, I really had that yokel's eyes popping."

"Yes," I said, "and it cost you about three dollars to do it. He already had his time card in."

But Randall never wasted a dime of the company's money. He and Joe Hachey could get more out of each piece of lumber and pound of nails that anyone else.

During the job we had to do considerable trucking on some unpaved gravel roads and use quantities of calcium chloride to keep the dust down and, of course, the temperatures of the house occupants along the way. I was to learn then that advertising is not always truthful.

The bridge at Pleasant Street got an award for its beauty and utility, and a photograph was taken showing the concrete and the bridge, and the concrete pavement on the Turnpike when under it.

The chemical company used this photo in its advertising, with a caption that said, "Cured with CaC12, Calcium Chloride." Actually, no calcium chloride was used in the concrete on either bridge or the highway. I wrote the chemical company, but needless to say, I never received an answer.

About this time, we got a big job on the main dam with the Metropolitan District Commission (MDC), working on the Quabbin Reservoir. Ed Kelly was to be superintendent. I asked Jack Doherty, Perini's General Superintendent, if I could be assigned there, because I had been in Quabbin before.

To show how people can find out things, one of the truck drivers, H. Callahan, told me, "You won't go to Quabbin. You're going to run Fred's job in Podunk."

The general superintendent did not know this, but sure enough, Louie Perini came to me and told me his brother, Fred, who had broken away from the family years ago and was not very successful on this own, had a job to build a road and couple of small bridges between E. Brookfield and Charlton, through a settlement called Podunk. The road was called Podunk Road. Louie said the job was in a mess and the company had their own equipment and money in the job and hoped to at least not get stuck for any more.

I went to Podunk the next Sunday and drove through the site. It was in awful shape, and I thought of quitting, but it was in the mid-1930, during the deep Depression, and jobs were hard to come by. I went to work at Podunk.

The job at Project P as we called Fred's job in Podunk, was really in a mess. Fred had an ancient steam shovel that he loved. He had been a fireman and operated on the old rig when he was a boy working for this father, Bonfiglio, who had started the business. The shovel, a real antique, could not produce enough to pay for the crew, coal, and steam cylinder oil.

The first thing I did was to park the old shovel and con Louie Perini into sending a modern gasoline-powered rig to get the job going. I also talked Louie into sending me Joe Hachey to be foreman on the two small bridges. Things got better, but Fred's credit was still poor. It was hard to get gasoline.

One day, when we were almost out of fuel, I got ahold of my friend, Jonesy, who had a gasoline distribution outfit and had supplied us at the Quabbin job. I got him to bring over some gas to fill up the tank. I guaranteed him I would pay for it myself, because he told me he knew that Perini Co. would not pay any of Fred's bills. I had a hard time collecting my wages from him, and his brothers finally paid me.

Anyway, he sent a tanker, and we filled up the tanks. Just as the timekeeper who distributed the gas to the shovels, bulldozers, and graders was about to fill a 55-gallon drum in his pickup, Fred arrived. There was a minute leak near the glass bowl of the pump.

Fred told the mechanic, "See that leak?" (It might have been a pint a month.) "That's how we lose money. Give me a wrench." He took the pump apart and managed to break the impeller. The nearest available spare part was about sixty miles away. So, we had no gasoline.

I called Jonesy and told him what'd happened and asked him if he would send a tanker truck over to fill up our equipment in the field, so we wouldn't have to shut down for the day. But he really let me have it, telling me I was worse than Fred and was trying to put him in a worse hole, so no gas that day and no work to be done. When the part arrived, Fred put it together and was very happy, because he had fixed the leak.

There was a shortage of gravel in the area, and it was not gravel country. Fred had bought a piece of land, cleared, and stripped it, preparatory to graveling the road. He had not dug enough test pits, and when we opened the pit up, we ran out of gravel in a very few days.

I went looking for new pits. One evening, Fred joined me. One place looked favorable, but when I climbed the hill, I found a small, shallow pond, and I told Fred that water could not be there unless there was a thick layer of impervious material and possibly no gravel at all. Later, I found some gravel, and we solved that problem.

Fred didn't know anything about bridges, and I told him to leave Joe Hachey alone, but Fred came by one day, when the carpenters were having a tough time getting the footing forms placed on the edge of the stream. Fred started to give Joe a lot of orders. Joe was a gentleman, but he did not suffer fools gladly. He came out of this hole with a hammer in his hand.

Fred fled and told me afterward, "I only wanted to help him."

The next job I went to was Cape Cod Canal. The Corps of Engineers had finished building two high-level highway bridges and one vertical lift railroad bridge over the canal, and we had contracts for excavating to widen the banks. Our contract called for excavation for five feet below sea level and placing slope protection with crushed stone. The berm left along the canal was to be removed by dredge, which would be done by Great Lakes Dredging Co.

In order to excavate this land, of course, we had to dewater the excavation down to five feet below sea level, and the depth of water would be more than this because of the tides. The company made a large investment in Moretrench pumps and well points. This turned out unsatisfactory, because the canal was so close that, actually, we were trying to pump out the Atlantic Ocean. However, we solved this by sectioning off the area and driving some wood sheeting for cut off, doing a relatively small area at a time. Then, we graded the slope and placed, with scale pans and cranes, the broken stones and riprap for slope protection.

In my section, we had to blast and remove a series of concrete piers on the land and an abutment of the old canal bridge. This blasting had to be done very carefully, as there was a high-tension power line running overhead. I did not trust the company blasters, so I did all this blasting myself. By this time, you may be sure, I had several more years of handling explosives than when I made my first mistake, fifteen years back. I was very proud of this, particularly when I had drilled and blasted the abutment and fired it so it slid into the canal. It slid in chunks into the water, and the dredge (thirteen-yard bucket crest), which is still in service, scooped up the pieces and loaded them into barges.

One sad thing happened on this job. Perini, on the south side of the canal, got all their stone by rail from the New Haven Trap Rock Co. in Connecticut, including the riprap. On the north side, my job, we got stone in hopper bottom cars, but we got our riprap from Acushnet Quarry.

When I had set up a loader in a pit under a spur track along the canal and was ready to place the stone and start placing riprap, I was visited by several representatives of the trucking companies and the quarry. Their direct questions were, how many tickets can you use each day?

Each truck carried a slip signed by a certified weigher, and we had a foreman sign the slip when the load was dumped into the scale pans. I told him that my foreman, who, like I did, knew such practices went on. On the highway jobs, when extra slips are signed each day, the quarry got a third, the trucker got a third, and the contractor a third, for so many tons of stone at so many dollars a ton. And of course, the Highway Department and, in this case, the Corps of Engineers were to pay for this thievery.

Anyway, I told them I would not have any extra slips signed, unless my foreman saw them dump and he signed them. In any case, if any truck was delayed and came in after the foreman had gone home, they were to wait until the next day and not come around with any tale of dumping after hours and expect to have the slip signed. They told me all the other contractors were doing so, but I told them we would not get into this scheme.

Then, I called Joe Perini, the treasurer, and told him, and he agreed, of course, with my attitude, reiterating Perini's policy, and told me to watch out. As a matter of fact, the Corps of Engineers got at least two loads that were really dumped after hours, but my foreman did not sign the tickets, so the quarry did not get paid.

We were buying our riprap from Connecticut by rail on the south side, and the bills were FOB, so it was to be paid by the Corps.

However, during the rail trip, even though the riprap was the correct size when loaded, there was some breaking and chipping, and the Corps inspectors were measuring up the load and deducting it from our pay.

Arbitration was set up, and it was agreed that Mr. McLean, the Deputy Chief Engineer for the Corps, and I would travel to Branford, Connecticut to watch the cars loaded and then watch them unload for a day or so, to see that the quarry loaded the proper size stone and that it really got broken down in transit. We found that we were right, and McLean and the Corps agreed not to deduct the small pieces anymore. We used it in the broken stone.

On the road back from Connecticut, we stopped for gasoline at a garage between Acushnet and Cape Cod. There were two big Sterling trucks there, owned by one of the trucking contractors.

McLean asked the proprietor how long the trucks had been there. The man said they had been there a long time, since they had been repaired. He wished the owner would come and pay him and take the trucks off his property. This, of course, intrigued McLean, because it had always appeared these contractors could use more trucks.

He asked me why our stone had always been close to the theoretical quantities and the other contractors were much higher. I told him I thought that they might have had softer bottom than we had.

He said, "No, I don't think so. The bottom is about the same everyplace. You have soft bottom here and there, too."

So, anyway, we took down the trucks' registration numbers and went to the Corps office at the canal. An investigation of tickets they had been sent for payment showed that these two trucks had been getting slips signed several times a day during the period that they were in for repairs. There was a big investigation, and the contractors had to give back some money. It was not really enough, and they sent one man to jail. He was a low-paid weigher at the quarry. Of course, he did wrong, but the others who really made the money out of it were all acquitted.

The next move was to take some time from the canal job and go estimate several jobs our company was going to try to bid on. There was a large job on the Newburyport Turnpike, so they put a substitute in for me at the canal, and I went off to figure the job.

On the turnpike, there were thousands of yards of gravel required. To be low bidder meant you had to find a place near the highway where this material could be found at a low price. They found such a place, and it belonged to the Boy Scouts of America.

I called on Charlie Pinkham, the nephew of the famous Lydia Pinkham. He was head of the Scout Council and had been the Republican candidate for Congress in the area. As an aside, he told me that he had never thought he was a saint, but until he got in the way of the mudslinging of the campaign, he did not know what an evil devil he was. Anyhow, we made a successful deal, and for the price of the gravel, we built a lake, beaches, playing fields, and campgrounds, water and sewerage systems. All came out fine.

When I finished my reconnaissance at this site, Louis Perini, Joe Perini, Jack Doherty, and Al Latanzi came out on the site. I had a chauffeur, a driver who was a truck driver. Louis had a chauffeur, a Narragansett Indian, very dark and fierce-looking. And each of us had a black sedan—Buick, Lincoln, Cadillac, and so forth. We met at Howard Johnson's.

We left all the cars but mine, and I drove the five of us around the area, pointing out the good and bad and so forth. When we came back to the restaurant, we found the place surrounded by State Troopers and the chauffeurs undergoing a great third-degree. They had gone in the restaurant for coffee, to wait for us. The proprietor, suspicious it was some sort of gangland rendezvous, had called the police. Of course, it got straightened out, but the drivers had had some very anxious moments.

They had bid another job at the same time. I had not done any work on this job. It was a grade separation job at Barber's Crossing in Worcester. There was a highway bridge, a railroad bridge, and a tunnel for pedestrians under the railroad. The job required about

150,000 yards of embankment. The estimators working on the bid had not found any source of fill nearer than a couple of miles.

The day before the bids were due, Louie said he was not really satisfied with the price we had for the fill material. As I completed my work on the other job, he wanted me to run up to Worcester, about twenty miles away, and see if I could find suitable material closer to the job. I had seen no contract documents. All I knew was that the job was a railroad crossing on the road from Worcester to West Boylston.

I arrived and found a hill near the Worcester Press Steel Company, close to the crossing. The material was clay in appearance. There was no time to dig test pits. I met Mr. John Higgins, a courtly gentleman, who was president of the steel company. He had been visited by our chief competitor in the bidding but had made no agreement.

We looked at the area, and he showed me where he wished to excavate an addition for his building. I agreed to do the excavation for the job for a very low price, if I could use the fill and dig further into the adjoining hill and pay him a price per cubic yard for the material. By using this pit, we reduced the haul by over two miles. This of course allowed a reasonable reduction in our price.

Louie promptly reduced the bid price by a small amount. I told him that our competitor, who had already located the pressed steel pit, would probably want to bid a little lower than that. However, he said that, all in all, we had a good price for the job. He thought we were low enough.

The next day, when the bids came in, I found out that Louie had had second thoughts that night, got out of bed, and cut the price for the borrow material by several cents, according to my recommendation. When the results came in, we found we were low bidder by a small amount, but Louie's cut off of about $10,000 was unnecessary. We would have been low by $3,000 anyway. This $13,000 was not, of course, a very high percentage of the total, but Louie was irritated and sent me up to the job with instructions that

I must make up the $13,000 in savings. We did make it up, and we made a good profit.

We started the job in the fall of 1936, and I moved up from Cape Cod. My timekeeper was a man named McGillicuddy. He was new to me, although he had been with the firm before. He was to work with me from 1936 through 1940 on several jobs. I brought my carpenter foreman, Joe Hachey, and road foremen, Angelo Brovelli and Frank Liana. The resident engineer from the State was Andy Paton.

The Worcester job was composed of over 200 items, from clearing, stripping, paving, concrete, steel, and so forth. As with many jobs, the land taking or rather, with this case, the building taking, sometimes causes delays. The construction was delayed, and we were forced into winter operations. It turned out to be a mild winter, and our one boiler, which heated the water for the concrete and supplied the steam to heat the concrete and forms, was inside a covering we built of timber and heavy building paper.

The boiler was idle for many days, because we had an extended January thaw. We had weeks up to 65°F, and we could play golf. We had to protect all the concrete abutments, the highway bridge, and the railroad bridge. We never had to heat the concrete decks on the steel girders. We didn't place this concrete until mid-April, but oddly enough, there was a freak snowstorm, and Andy Paton came out himself and covered the freshly placed deck concrete with canvas to protect it.

We had planned to complete the paving and build the underpass under the railroad in the summer of 1937, but the city did not get the buses to replace the streetcar line, which crossed the railroad at the joint crossing. And so, there was another hiatus, and we ended up doing the tunnel work in cold weather in early 1938. During this delay, we did not wish to move out and shut the job down, so I got some more work in a nearby city, flood control, to use some of the equipment and also, to some annoyance of Andy Paton, I picked up several small excavating jobs to collect additional fill to complete the highway embankment.

Andy had to get his survey party out to cross-section it. I also built an access road and graded a playing field for the Norton Company, a large manufacturing plant nearby. The city finally got the buses so that the trolley tracks could be removed.

The design called for piles to be driven on the area of the railroad crossing, which would be capped with timber, and the track would be supported with it while we dug under and placed the concrete for the tunnel. The tunnel had mechanical sump pumps and lighting and, of course, electrical gear to go with it.

We drove a test pile alongside the track, so we could estimate the length of the piles required. This pile went down easily, and we ordered piles of that approximate length. When the piles arrived on a flat car, we proposed to drive them with a railroad steam pile driver. We found that all under the tracks there was a glacially transported mass of boulders, and the piles could not get down more than two feet or so.

We dug around the test pile and found that there was an old sewer pipe passing there, the existence of which we had not known, and neither did the sewer department. The test pile had been driven in the backfill on the pipe adjacent to the edge of the old brick sewer, and two feet one way would have hit the sewer, while two feet the other way would hit the mass of boulders. We had to cut the piles into short lengths and excavate and post down to hold up the tracks. It was a tedious operation, but we had bid a token price for unclassified excavation.

The railroad had an old foreman there named George (Jarge). He was very helpful and gladly took our gasoline-operated concrete vibrator in the shack there, to keep it warm. Except he drained the gasoline on the ground outside his shack and threw a match into it, so it would be gasless when it was put inside. He almost lost his life, the shack, and our vibrator.

He used to talk about his brother down in Maine. "He's dead. He got caught between two trucks," he'd say, and shaking his head, "Couldn't stand the pressure."

We had other small problems. There were two tunnel entrance buildings, brick and concrete with slate roofs. The contract read "entrance buildings (2) lump sum." Our estimators had figured the cost of one building and did not multiply by two.

When I figured the amount of brick to order, I found that the bricks and laying would cost more than what we were getting for the lump sum. I called the main office, and Louie said, okay, we made a mistake. Just get the loss over as quickly as possible.

A similar thing happened with the electrical work. As is common, estimators work on sections, and all the figures are put together and extended. The electrical estimator had figured the labor down to the last penny but omitted the value of the materials. We had to take a small bath on that one. We saved a little on the electrical boxes, because the specs called for cast iron made by Russel Stoll, catalog #39A, with lock and key. However, the Russel Stoll boxes did not have any lock and key, and to fit, this would require new patterns and special flasks and so forth. We were allowed to build them out of aluminum, which saved us a little.

This contract had been designed by the city, let out to various engineers, and the state was going to supervise the construction, while the federal government would pay for it. Beside the electrical problem, we had a problem with the contractor about the brick work.

We didn't mind the spec for Flemish bond, although that's more expensive bond to lay, because you have to cut bricks in each corner, but the specs called for Antique brick. Antique bricks are malformed, overburned bricks from being close to the fire. They give a pleasant texture, with a few of them staggered around the wall, but to build a whole wall is not only costly but certainly does not give a pleasing appearance.

We arranged to get some hard-burned brick to make the architect happy. The structural engineer had designed the drainage manholes and the sumps. This design looked pretty, but it was impossible to build. Paton said to call the local engineer.

This gentleman listened to me and said, "That's your problem," and hung up. Andy and I got together and redesigned a method that was workable.

This job was the first union job Perini had. The laborer's union was not well organized, and they came working from the state employment service, anyway. The ironworkers, carpenters, and bricklayers all came around and made their pitch. Benny Gordon of the carpenters came. "All we have to sell is service," and he delivered the service.

The rate for carpenters was a $1.20 or $1. The rates in the contract for union carpenters furnished by the union would be less than the rates for carpenters furnished by the employment office. The union supplied a certain number of lead carpenters at $1.20 and the majority at a $1.00, so we benefited by using the union.

Perini had always bid ironwork at so much a ton. Pug Reardon, business agent for the ironworkers, said his men would save us money, so we went union ironworkers. Louie came up one day and watched these men work then said he had never seen better men on both reinforcing and structural steel.

Henry Brennan, who later became vice president of the International Union of Bricklayers, came in, and we made a deal with him on the brick laying, cement finishing, and so forth, which we never regretted. We did have a little trouble here because the tunnel walls were faced with a veos tile, which the specs said was enamel-faced metal tile, which is usually laid or set by carpenters into a grooved wallboard and glued into place. These tunnel walls were designed to be concrete, lined with split-ribbed tile for drainage and faced with metal tile set in 1-inch mortar on the split-ribbed tile.

There was an ongoing dispute in Boston between the carpenters and the tile setters, who were part of the bricklayer's union, over which trade the material belonged to. The subcontractor we had hired used carpenters on the wallboard setup. However, we all recognized that the carpenters could not set the tile in mortar. Brennan came to the rescue. He agreed that the subcontractor's foreman could supervise and even cut the tile with a power saw, and his people could work on my payroll, and we could backcharge the subcontractor.

I was later, in 1942, to invoke Henry Brennan's name to my advantage in upstate New York. As usual, I did not get to finish this job. There was nothing much left except for some erection of chain link fence; my old friend and fellow worker, Soma, who had taken over the Cape Cod job, came up to finish this job.

I went to Union City to work on the Lincoln Tunnel between New York and New Jersey. One reason I think was because Louie Perini felt I was the only one of his superintendents who had any experience with unions. I had always been a strong advocate of unions and had some brief experience with carrying a card and working as an ironworker.

There had been, over the years certain excesses and corruption in the operation of the unions. It was not long after this that Fay and Bove went to prison for corruption in New York and New Jersey. (Joe Fay was in the Operating Engineers and Bove in the Labor). But I remember well how the employers treated the employees before they organized. In 1937, originally, the laborers might be getting forty to fifty cents per hour, and the job engineer would have $60 per week. Today, the laborer's rate is over ten times that, but the unorganized people, whose rate can be set by management, do not receive ten times by any means.

The one outstanding thing that happened to me at Barber's Crossing was meeting and working with Andy Paton. We did that job, which had over 200 items of work, and when final payment was offered, our books were only $700 out of a several-million-dollar project. This, I remember, was for topsoil stripped and stocked and rehandled.

The specs said there was no need to strip loam below more than four feet in depth of fill. We had stripped an area and saved the topsoil to place on the slopes. The finals department had deducted the $700 from our requisition for this reason. However, this $700 was actually less than if the state had to pay us for bringing in loam from outside at a higher price. Paton told the Highway Department the story, and we got paid.

There are two other stories among the many Andy and I could swap. We had a dual-drum paver, and although Perini had an excellent operator, Vic Riga, who would keep a practically steady stream of concrete flowing, Perini wanted to keep him on another job, where he was rock drilling superintendent. Instead, they sent old Joe Zafferano. He weighed about three hundred pounds and was also called the Sea Pig. Joe was a fine roller operator and a good man on a single-drum paver. The dual-drum job was too much for him, but Louie did not wish to change.

Then, one day, when we were pouring concrete on the railroad bridge abutment, Louie came to visit. I repeated my wish for Vic Riga; however, he wouldn't even discuss it. While Louie, Andy, and I watched, Joe the Sea Pig was struggling with the machine. The foreman, Angelo Brovelli, according to his usual habit, threw his hat on the ground and jumped on it. Louie asked me what was the matter.

I said, "He hasn't even got a batch in the skip."

I went to Joe and told him that. He raised the skip and started the mix again. The next morning, I had Vic Riga on the job.

The other thing Andy and I remembered was the bridge rail on the highway bridge. The architect had designed a fine-looking bridge fence, all concrete with panels about four inches thick that would be bush hammered to expose an attractive mix of colored aggregate.

Because of the change in texture, it had to be carefully placed so as not to get any rat holes or honey combing. When one section was stripped, the wall was not looking too good. I came along and joined a group of inspectors, cement finishers, foremen, and so forth, who were discussing means and methods to repair the mess.

Andy still remembers me telling a worker, "Hand me a sledge," and then taking the sixteen-pound hammer and smashing a section of the wall, then telling the foreman to rebuild the whole thing. It looked beautiful when it was rebuilt. But push hammering removes a film of protective cement, and I feel the fence did not keep its beauty for very long.

We found in Union City that we were faced with different problems, as usual. We had to cut through between 34th and 35th Street in Union City. All the buildings had been torn down, and rock had to be removed between these two streets to considerable depth for the sunken highway approaches to the Lincoln Tunnel. We had to relocate utilities, build a highway bridge to handle a traffic trestle for the street railway, build trestles for sewer lines, relocate ADT telephone and water lines, encase storm sewers and sanitary sewers, and build reinforced concrete retaining walls around the three-level structure, which was to carry the Hudson County Boulevard traffic over the main roadway. All this required design on the spot.

This was part of my job as engineer. Ed Kelly was super and handled most of the rock excavation. I designed a bypass, the first thing to go around the future structure to carry the Hudson County Boulevard traffic. This was to be a steel-and-timber structure.

When I finished the work, I sent John Gibbons, our office manager, down to the Building Commissioner's Office, Hudson County Boulevard Commission. He came back pretty well shaken. It turned out the HCBC was a political front, and the Hudson County Engineer, Frank Radigan, was the man. When he was checking the drawings, he complained there were no blueprints, the same complaint I'd had five years earlier from the Boston & Maine engineers at Haverhill, New Hampshire, and that Gibbons was not an engineer. Mr. Radigan wanted to talk to me.

He was a common-sense gentleman; he readily waived the blueprint requirement and approved the drawings. We had some conversation about Perini's experience and mine with construction. He seemed a little bitter about one of our competitors, who was doing some work for Hudson County near our job. It appeared that their superintendent had crossed swords with Mr. Radigan, and he wouldn't give Brewster the time of day. Brewster was also a Republican, and that might have had something to do with it.

Anyway, I didn't see Mr. Radigan very much. I took him to lunch a couple of times, I gave him a cigar another time, and we stayed friends. Later, we had to move a large Caterpillar-mounted crane and large bulldozer. This should have required permits and engaging Gerosa, the company in the Bronx, to move the equipment. A telephone call to Mr. Radigan and a $5 fee to the New Jersey Highway Department were all we needed to move both pieces of equipment along the County highway.

We had to lay planks on the pavement under the bulldozer. Mr. Radigan also sent some Hudson County police to care for the traffic. This only cost us some cigars for the policemen. Later, when I wished to get a P.E. (Professional Engineer) license in New Jersey, Mr. Radigan was one of my sponsors. He had been gone for many years, of course, but he was helpful; even after World War II, there was a recommendation for me, "a fine man." Hagne was mayor of Jersey City then, and of course, Radigan must have been associated with that group, but you would not have thought he was any part of any corruption or believed it, if you had been as close to him as I was.

When we uncovered the storm drain and sanitary sewer pipes, we had to encase the concrete under the fill for the ramps. We found out that instead of being side by side, as shown on the drawings, the contractor had evidently laid the 12-inch sanitary line underneath, as an underdrain for the 48-inch sewer, which was lying practically on top of it. This gave us a bit of trouble to redesign the encasement.

We also found that the contractor had decided to discharge both lines into a manhole that led to a trunk sewer heading to the North Bergen Sewage Treatment Plant. I asked Josh Muss, the plant manager, if he had any excess water coming into his plant when it rained. He said it was much more than he expected from ordinary infiltration, and he could not treat it and had to bypass it. We didn't fix this, only encased it under our abutment, but I believe that Muss's people got the drainage straightened out.

We had many problems with the delivery of steel because, in some areas, you could have steel bent by the factory, but in New Jersey, this local, and the union local in New York City said the steel had to be bent and cut on the job. We had to get that straightened out. The road went through an old amusement park, and there was swimming pool equipment and large steel filter tanks filled with sand that were the property of the owner of that pool and park. He was supposed to clean the sand out and take away the tanks. However, we were in a hurry and proposed to move the tanks to one side. But we couldn't get the ironworker foreman or the operating engineers to hook up a sling on the tanks so the shovel could lift them out of the way.

The union, then and now in most places, requires five men for a crane: pusher, signal man, tag man, etc. I refused to hire these people, and of course no laborer or oiler would hook up the slings. Neither would I. I took two bulldozers and pushed the tanks away, and the five union workers walked off. I knew times were hard, and they needed the money, but so did I.

We had our office in an old brick factory building and our air compressors in the same building. These compressors were operated by electricity. The Teamsters and riggers deposited them on the sidewalk and went away. Then, ironworkers or heavy-duty mechanics with jacks and blocks moved them over the anchor bolts of the foundation, which had been set by carpenters and millwrights, who are part of the carpenters' union, and lined them up and bolted them down. Then, we had to have an operating engineer be there during working hours. We had to put a little oil here and there, but mostly nothing.

Local 3 of the IBEW, International Brotherhood of Electrical Workers, a very powerful union then (and still is; probably more powerful now) wanted to have an electrician there at all times. We argued that there was nothing to do. We finally came to an agreement that the local electrical subcontractor, who had union men at work who had little to do with us except to set junction boxes in the bridge abutments, would be alerted if a fuse had to be changed. The office manager would call the electrical sub and tell him, then the timekeeper or one of the clerks would change the fuse. An electrical sub would pay at half day some electrical union member who was sick or disabled and bill Perini for the cost.

The compressor operator complained about the heat and the enclosure. He was told he could put in a window, if he wished. He took a hammer and chisel and knocked out some brick in the wall. The next day, he was waited upon by the Laborer's business agent, who made him pay a laborer a half-day's work for taking his rightful work away.

To handle the ADT fire alarm and the Hackensack water lines, we had proposed to put them under the sidewalks on the boulevards temporary bridge. The water pipe was only ten inches of cast iron. The manager of the company, Mr. Griterage, was really cautious and came to see me to check if the bridge deck, which was designed to carry a ten-ton truck, could properly support his water pipe.

A telephone line had to be moved, and we had to locate it; then, the telephone company was going to do the work. At the Hudson County boulevard, there was a dual carriageway, and the drawing showed that the telephone cable was in one of the carriageways, the eastern carriageway, so we started to dig in the western carriageway. We put a man with a pick and shovel to dig a test pit where they thought it was in the eastern carriageway.

Patsy Susi was an old laborer who had been with us a long time. The first thing that happened was that he dug a hole here and a hole there. There were about seven people there: people from Perini, the Port Authority, and the telephone company. Finally, Patsy threw his shovel and pick down and said, "I quit. One man at work and seven men to watch," and he walked away.

About that time, no one had to do anymore test pits, because the shovel that was digging on the western carriageway found the cable and tore it up. There were about 200 telephone cars there in five minutes. It took a little time to get some of the telephone service back in shape.

The trolley line at Summit Avenue was supposed to be carried in a trestle over the excavation. Here again, I had to design a trestle for a temporary crossing before the final bridge could be built. Of course, the street railway people went into the usual ploy, demanding blue prints instead of white prints. I had laid out curves that were identical to what they had used in some of their turnouts.

What finally happened was that they never put their cars on the trestle, and it ended up being a trestle for cars, trucks, and pedestrians. The trolley cars never had to cross, and the customers walked across. Of course, street railways were going out of business anyhow, and I do not remember if they ever put their cars in a permanent trestle when it was built. As I, of course, left the job before the thing was finished.

There was a small misunderstanding in the street railway business in setting the salvage value of the track rail and so forth. I brought it up to Louie, we had not figured this little item closely, and it appeared that it might be a $7,000 difference in the interpretation on the contract. The street railway people did not care; it was between the contractor, which was us, and the Port Authority.

I brought Louie Perini to the office of the chief engineer of the Authority, Colonel Charles Gleim. Louis stated the question: we would like to get the $7,000, but we would accept the decision based upon the intent of the people who drew up the paper.

The Colonel said, fine, and called in the engineer who had written the spec. He asked the man, "What did you mean by this paragraph on page 23?"

The man said, "Of course, just what it says."

The Colonel said, "Okay, look at page 92. What does that mean?"

The man started again, "Well..."

Col. Gleim said, "These people recognize the difference but are willing to settle for whatever your intent was in the matter."

The old engineer admitted he did not know what his intent was. So, Gleim dismissed him and told us to get out of there and do whatever we wished. And we did.

We had a great office manager, John Gibbons, a real sharp office man, philosopher, and a wheeler-dealer. He used to get all his work done by 11:00 a.m. He always said that no man had an original idea after 2:00 p.m. By noon, he would organize a lunch date with one of the suppliers, say Eddie Mahoney of Mahoney and Clark.

Gibbons would ask me, for example, if we needed any air hose, and if I said yes, we could use some in a few weeks, he would call Eddie and tell him we needed it right away. Eddie would put it in his Cadillac and drive over and deliver it, then take Gibbons to lunch. After lunch, he was usually at his desk, doodling, his day's work done.

He organized a club called the Four Thirty Club. The shift changed at 4:00 p.m. This club headquarters was a bar and grill, Louise's Tap. Louise was a strong Italian woman and her husband, Vic, was the bartender. All of Perini's supers and engineers and inspectors and so forth plus officials of the Port Authority would figure their daily rounds to arrive at Louise's at that time, to eat and drink, mostly drink, until it was time to go home for dinner. It was a fine social custom. Gibbons would go back there the next morning to get a pick-me-up and pay the tab for the day before.

Bartender Vic had been a rum runner in Amsterdam, New York. His favorite expression was "on the house," then he would collect the money all the same.

In those years, 1938, business was so quiet, we could drive through the one lane that was open at Lincoln Tunnel, get on 42nd Street, and park at the ramp door of the Commodore then go into dinner without any bother about parking regulations. At the dinners, Gibbons would entertain with fine Irish ballads.

Eddie Mahoney would be there, along with Ed Kelly, Jack Doherty, and Louie Perini sometimes, if they were in town.

Gibbons would sing, "My name is Kelly, Edward Kelly, but I'm living the life of Riley just the same. I'm drinking Riley's booze, I'm wearing Riley's shoes. The only thing I haven't got is Riley's Irish nose."

Then Ferguson, an engineer from the Port Authority, would recite his poem. "There's a green-eyed, yellow idol to the north of Kathmandu. There's a little marble cross below the town. There's a broken-hearted woman, tends the grave of Mad Carew, and the little god keeps ever gazing down. He was known as Mad Carew by the subs at the Kathmandu, but for all his madcap pranks, he was worshipped in the ranks, and the colonel's daughter smiled at him, as well."

Ferguson had been a Marine in Haiti and could tell some fearsome tales of burying unruly natives on the beach with just their heads showing and leaving them for the incoming tide. But "Mad Carew" was Fergie's main contribution to these discussions.

My next move was in 1938, to Pittsburg, New Hampshire, to build a dam for the New Hampshire Water Resources Commission. Governor Murphy had organized this commission for the benefit of the state and for the benefit of the New England Power Company, so that land could be taken by eminent domain. This reservoir dam could be built on the Connecticut River.

Pittsburg is practically all of New Hampshire above the 45th parallel, across to Maine and topped by Quebec. The power company had two dams above Pittsburg and three dams now impound all the water up to the Canadian border.

After the American Revolution, Vermont was not one of the thirteen colonies. After a while of staying independent or joining with Canada, Pittsburg became, at that time, the Indian Stream Republic, with a standing army of over a hundred men armed with muskets and pitchforks. But the New Hampshire governor sent forces up there and annexed the place.

When I was there in 1938-39, the people were still fiercely independent. A contractor from Massachusetts had moved in to build a highway several years before, and he would not hire local people, but brought in all his own men. Someone put a case of dynamite on the floor of his power shovel and blew it to pieces.

Because of this information, my first move was to hire a personnel director locally, set a small office (tar paper shack), and start hiring. Until Mr. Putney had hired all the men from Pittsburg and Clarksville who wanted to go to work and had sharp axes, we looked no further for that type of crew.

The nearest railroad was ten miles away, at Beacher Falls, Vermont, and all the equipment came in there. Some of it walked along the roads and forded a few brooks before it got to the site. Charles T. Main Inc. had designed this project, and Billy Uhl, who was the president of the firm at that time, was a frequent visitor as was the manager of the power company. Colonel Jacobsen was the head of the Water Resources, and Dick Holmgren was chief engineer. Holmgren was a Quaker, and Jake thought it was amusing and used to tease him about it. John Baker, an old engineer, was a resident. I remember two of his wise sayings: "Even the mules know when it's a force account," and, "No man ever gets hurt on dangerous work."

We set up a construction camp complete with running spring water and a septic tank sewer system. Sam Golden, the operator of the Golden Commissary, who had operated camps before for Perini at Enfield, set it up. He called the workmen who lived and ate at his camps his clients. Sam's system was to feed them the best steaks at first, to gain their confidence, then he would taper off.

He had a knack for looking ahead, so when the rolls started to become depleted and the clients fewer, he would sell out to his partner and look for more golden pastures. On this job, he sold it to a man named Benny, who in time sold out to a man named Kelly. Strangely, I think Kelly made plenty of money, because he carried all kinds of clothes, boats, boots, and equipment.

When the camp was set up, we had used the US Army figures for wash basins, showers, and water closets, and built to Army standards. Sam said that we had too many showers, because these people wouldn't take them. Of course, this was true for the lumberjacks. But later, after the site was cleared, operators, iron workers, carpenters, and so forth came from outside, and we had the right proportion.

The dam work went reasonably well, delayed by problems of getting reinforcing steel, so we had to work all winter to get the stream diverted before the spring floods. This meant three boilers: one to heat the aggregate/sand and gravel storage bins; the second one to heat the water for the concrete; and the third one to make steam for unit heaters in the formwork. We had a covered enclosure to keep the concrete warm and to clear the ice off the rock, when we moved the enclosure.

We brought considerable soft coal, which was then about $8.50 a ton with only $2.40 run of the mine; the rest was freight. Because there was no suitable hard clean sand and gravel in the area, we had to come in by rail via Beacher Falls. The winter of 1938-39 was really cold; it got down to -40°F and the snowfall reached four to five feet. It became difficult to get the frozen sand out of the bottom dump cars. We used to help this with a few sticks of dynamite. Once, the foreman set up too large a charge, and it broke the springs of the car, which was of course brittle in the low temperature. We had to splice it with timber, and then the railroad hauled it off to their shop for repairs.

We started the work double-shift because of the time factor; that meant we were working seven days a week, ten hours a day each shift. In Beacher Falls, there was a very fine middle-aged, Catholic priest, a Frenchman from Quebec, and quite a few French people, as well. He complained to me about our working on Sundays, and I said we were working two shifts and that, if he wanted his people to go to church, he should run two masses. So, he did that to keep his business.

He also complained to me that these people were making money for the first time in years and weren't paying any dues to the church, whereas the church, like all the churches in Quebec, built a town around the church. When the people don't have any money, the church supports them. This priest discussed it with me, and then he said what they do is go to the barrooms all night, but don't put any money in the church. Then he got a good idea. He took a big can and used to go to each of the barrooms, where he collected money from them to run the church.

We had a hurricane in 1938. We weren't getting very good weather reports, so we didn't expect it. We were struggling with the Connecticut River in flood. I always say, when you design a coffer dam and put it into service, it almost always happens that either it is a drier season than ever on record and your coffer dam looms up like a sore thumb or the thousand-year flood takes place and you are putting sandbags on the dyke and looking for fat Dutch boys to put their thumbs in the holes of the dyke.

The flood started to subside about the same time as the hurricane flew in at about 2:00 a.m. I ordered the pumps pulled out to hold the dyke by letting the water build up inside. By this time, our towers and lights had blown down, and my master mechanic and I had pulled each other out of the water about five or six times. The hurricane did lots of damage to the trees, because the direction was 180 degrees from the prevailing winds in that area, and the trees were not rooted in against that direction. By the next day, we managed to get some poles up and pumps back in again, and we didn't lose too much time about it.

As the dam proceeded, we had to clear the flowage upstream in the dam. This area was flowage where the reservoir was going to be setting and was about seven miles long. We had to log out the timber. The good-size timber we logged out was suitable, and we sawed it up for our own use then sold the rest. For the pulp, we stripped off the bark and cut it in lengths for pulp wood to sell.

One failure we had was with the junk trees, like the piss elm, which no one used for saw timber. Someone, who shall remain nameless, from the power company said we could cut the elm down and cable them to the stumps, then they would stay there and gradually rot away as the water rose in the reservoir.

Well, there weren't very many of them, but later, during the hurricane, one of our foremen, Jack Casey, who had been in charge of the job of building a log bridge across the river, above the dam, which the material and fill had to be hauled across. During the early stages of the floods, elms had broken loose from the stumps and floated down the river. Casey was like Horatius: he and his men were either along the bridge with pike poles, trying to fend the trees off from hitting the bridge. They were not too successful, and we had to rebuild the bridge later.

One winter day, when it was about 30 below zero, I was checking up on the flowage area. We supplied each pulp wood cutter with a horse and chain; he had his own buck saw. He would pull in the logs or twitch them in. *Twitch* means you had to drag it through the dirt. If you scoot it, set it, which means you put them on a sled to keep them off the dirt. In the case of this, the men handled their log up to their saw horse, sawed them up to a four-foot length, split them to six-eight-inch size, and piled the wood up in cords, 4x4x4. At that time, we paid $1 a cord, and a fair man could make $4 a day, which is about what he would make at forty cents an hour. Some good woodsmen could make about $8 a day.

On this day, I saw a little Black man with practically an icicle hanging out of his nose, sawing away. He did not work for me, but several groups had set up camps and lived in the area. He told me he had been a bellhop at a local hotel, a big resort hotel in the White Mountains, and got drunk and lost all his money in a poker game the night before the job ended. He was earning his way back to Alabama. He wasn't making it very fast; he barely made $2 a day, mostly less, and paid eighty cents for food.

"If I can get out of this here, boss, I'll never come this far north again."

I saw him on the road several weeks later, walking toward the railroad station about eleven miles away with his buck saw wrapped up in burlap and his duffel bag over his shoulder.

The camp cooks are a rare breed, mostly anywhere and mostly alike. The heat from the stove or the exhaust and drinking their extract of vanilla or things like that. There was a cook at the camp in Enfield who drove all the clients out in the snow one winter's night with a cleaver. This cook in Pittsburg, Levi Gardner, did not drink. He was a stickler for form. No one could get into his mess hall until he yelled, "Come and get it!"

He was a fast worker; we would get all his stuff to make pies, for instance, and then stand there while the workers handed him his things, and he would literally make dozens of pies in an hour and stick them in the oven. He had to have hardwood chunks to burn. I asked him once if I could get him some anthracite coal with the steam coal to help him cook. He said he would quit; he could only cook with wood.

One day he did quit. He told me he was giving it up and going up in the flowage to cut pulp. I told him he had a room and a private bathroom, and he was getting $35/week and all his food and so forth.

"No," he said, "I can't stand that crazy Ernest." Ernest was one of his cookees. I told him to fire him, he works for you, you're the boss. But Levi went off to the flowage. He'd rather have his grievance—hot stove got him.

After the river was diverted, the pen stock completed, and so forth, there was some trouble with the inspectors at the tunnel job in Tuscarora Tunnel in Pennsylvania. So, they exchanged a couple of supers from there. I went down there with Al Lattanzi to get it straightened out. The Resident Engineer and his assistant and one of the chief inspectors had managed to get our three top men relieved. Details do not matter now, but Al and I managed to get all of them, three of those characters, fired from the Turnpike Authority eventually.

They say rock makes a tunnel man; ground makes a tunnel man. Hardrock tunnelers call the people *miners,* and sometimes, they called it *ground,* but soft ground, whether it needs air or not, is soft ground, and the people who handle it are usually called *sandhogs*. The rock in the east heading was the same chemical composition as the rock in the west heading, but the bedding was many degrees different and made a lot of difference.

We met some great people down there. I met Carl Rankin, a great tunnel man. We shook hands at the holing—the last shot was made through. He was a consultant for Perini on the job. Later, I was to compete in the pipeline business with his son, Clark Rankin, when he worked for Bechtel in Nebraska and Kansas. In 1964, I met Carl in Trinidad, where the contractors had brought him in to challenge me, so they could continue their sloppy work. Of course, Carl was not about to condone sloppy work, no matter who or how much they paid him.

Tunnel stiffs are, of course, a different breed. A powder monkey was called Hobo. Hobo called food *"chuck."* He would leave one boarding house for another because the "chuck" was better. He told me once a landlady served *"scrapo."* He said it was made of pig hair and cornmeal; I think he meant Philadelphia scrapple. We had a chuck tender killed, and Hobo said this was a great day for the stiffs. Because of the PWA policies, we had to hire some local people. This chuck tender had been a local boy. Hobo figured local boys would leave, and the stiffs would find out and more stiffs would come in.

Dirty Face Bill Jones was the super of the header where I was working. He was called that because chewing tobacco or snuff stained his face, along with the powder and dust. He had some quaint sayings. Someone jived him about the size of his penis in the shower room of the dry house. (Dry house is the locker room of hard rock tunnels. Hoghouse is the same thing in soft ground tunnels.)

Jones said, "I'm getting sick of hearing remarks like that. I'll either go back home where it won't be noticed, or I'll go to Hollywood where it will really be appreciated."

I still remember how Bill Jones was proud of his safety record and would talk of this constantly. At that time, there was much talk in the papers about deaths in the tunnels along the Turnpike. I think, in a couple of years of work there, about eleven or twelve died.

The day after the Sunday supplement was issued in the paper was the first day of the hunting season. At that time, in spite of the light brush and trees, they allowed hunters to use rifles on deer. On the first day, thirteen men were killed by hunters—all is relative.

We had many tunnel stiffs there. Reptile Red, Society Red, Breastboard Smitty, and so forth. Reptile Red had a handsome rattlesnake tattooed on his arm. Tunnel stiffs and pipeliners are not interchanged very often; they both stick pretty well to their own trade, and they seem to have similar type nicknames. I am very fortunate to have worked with both on tunnels and pipelines.

When I was still in the Army, running out my leave time after the war was over, I had a pair of compressor station jobs in Kansas, even though I was still in the Army. I came into one of our jobs at Mullinville, Kansas, on a pay day evening. The super, McNaughton, was in the office, along with Cherokee Marley, who was a splendid pipeman. He had been saved. Evidently, being saved meant going on the wagon, because blasphemy and violent behavior were not part of it.

Henry Fall Leaf also came in; he was a fine welder. He was working overtime to get his pay packet. As he opened it, he said something about the root of all evil.

I laughed and said, "You know, Henry, if you're talking according to the Bible, why it's the love of money that is the root of all evil."

He just laughed and went out. Then, Mack and Cherokee both turned to me and asked, "You read the Bible?"

"Yes," I said. "I've read probably most of the Bible."

Mack said, "I'm a member of Unity, and I went to Bible school."

Cherokee said, "I'm a Nazarene, and I went to Bible school, too."

Then, they started discussing verses in the Bible and very soon almost came to blows. The language they were using was profane and blasphemous, but they were both good pipeliners.

The pipeline business was new to me. I had really come to Kansas to see what it was all about, although it was many years before Perini took it up. I looked around for a few weeks, and

Charlie Foreman, president of the company, mainly of building pipelines and appurtenances for oil and gas companies and pipe for sewer and water tunnels, pipe and pumping stations, asked me to stay working for him.

When I said I had no experience, he said you built many things, "You know how to get it done. There's no difference here. What I will do is, I'll make you general superintendent and manager."

And he did. I stayed until I went into the Army in World War II. I came back from Italy just itching to get back into the pipeline business. However, the people from the East Coast are used to looking at regular streams; the drier muddy rivers brought me back to New England.

Back in 1946, I was approached by an old friend, Tony Cucolo of New York, a fine honest contractor, to join him and some others in setting up a new contracting company in New York City. Mayor LaGuardia was not in power; as one of the people said, "After reform, there is a rebirth of Tammany" or something similar.

Tony had a corporation in New York, Empire Contracting Company, which he had set up years ago with a proper charter, although he had as yet never operated it. He used to run down to Boston each week for a while. Besides Tony and me, the participants were Gene Pope, a big sand and gravel and concrete operator with the city, as well as the owner of an Italian language newspaper; Judge Mancuso; Hugo Rogers, president of the Borough of Manhattan; Mike Kennedy, Tammany leader; and Dan Scocozza, head of a big carting company. One of his connections, besides getting paid for the forty feet of snow, was to haul material into Madison Square Garden (soil, hay, and straw) for the circus and then clean up the Garden afterward, hauling the stuff to a point at W.130th St. and West Side Highway, where they'd store it for the next circus season. This was one of Consolidated Carting's annual jobs.

We knew he had connections. The object of setting up Empire Contracting Company was to get business from the city on roads, bridges, sewers, water, etc. I pointed out that the maximum contract that could be given without competitive bidding was $15,000, and we could not make much money on such projects. Then, I was asked if I could write the contracts and specifications, if we could be the low bidder all the time. I said I could be low bidder most of the time.

At this time, there were several large sewer jobs coming out, 72" and so forth, and I found that, under the new setup, I could get in the back way to the Chief Engineer's private office, have an office adjacent, and revise as necessary all the contract documents for these sewer jobs before they went out for bid.

About this time, Tony began to wonder if there were not too many people in this deal who knew nothing about construction. He decided to test the deal. We had lunch one day in New York City, and he said I should tell Dan Scocozza, whose office I was using at that time, that Tony only wanted to be a silent partner. He would put his share of the capital in my name. I was to do this and call him from a pay station in a parking garage.

I went to Dan's office, and he listened without a word. Then, he called Gene Pope and spoke to him in Italian. I had not been back long from Italy and understood Italian pretty well then. He was suspicious of Tony's methods. He said, "Jack, he's a great and brave man, but he couldn't understand what Tony was up to."

I was great, but that would leave only me as a construction stiff, and Tony would be in a difficult position.

Then, I asked Dan for the expense money I was getting from the deal from each participant in turn every week. He got a tin box out of his desk and peeled off several hundred-dollar bills and gave them to me. As I was leaving, I said something to him in Italian. He was stunned. He looked startled and asked me, in Italian, if I could speak it. I said, "Abastana bene," which means "well enough."

I called Tony and told him Dan had called "you know who" and spoke in the other language. After that, I went back to Boston. My wife and I were house hunting in the south shore of Massachusetts. We had dinner at Hugo's Restaurant, and when I paid the bill with one of the $100 bills, there was quite a commotion, checking a list of numbers, even making a phone call or two. However, the bill was accepted, and we had a good meal.

The next week was Good Friday, and I was with all these people on Good Friday morning. We ended up in an Italian restaurant in the Bronx, and then we were going to the opening day of the baseball season between the Yankees and the Senators. At this meeting, we had several bankers. I remember betting one of them that the Red Sox would finish above the Yankees that year. It was a small bet, but by the end of the season, I'm sure neither the banker nor myself remembered the other's name.

The game that day started at 3:00 p.m. I only had an attaché case with me with a change of shirt. During the game, I thought through all about the situation, and after the game, I excused myself and took a taxi to 125th St., where I caught the train to Boston. Tony agreed with me that there would probably be no new company.

There was a little action after Tony arrived, but things didn't work out. Tony and I still correspond. We were friends for many years, but we never did any business together. I never could tell whether, if I had gotten into this deal, I would be rich, dead, or in jail. Gene Pope died a few years later, and his son, Fortunato did not know what to do. He went to see Frankie Costello (Francisco Castillo). This I heard and knew no more about it.

Hugo Rogers dropped out as Borough President, and Kennedy was killed in an airplane crash from Boston to New York. Judge Mancuso was indicted for some other problem. Consolidated Carting is still in business as far as I know, and there is an MIT graduate called Gene Pope, Jr., Pope's son, who owns and operates the *National Enquirer*.

I think of Gene Pope, how when I was in his office one day, high up in Rockefeller Center, we had a similar conversation to the one with Dan Scocozza and a few hundred-dollar bills. He called Mike Kennedy, conversation in English of course.

Then, he put the phone down and said, "Excuse me. That was Jack (Tammany leader)." I was impressed that he wanted to impress me, because I knew he was a big shot, anyway.

In 1932, Perini had bought concrete aggregate from Gene's Colonial Sand & Gravel. One of our plants was at Grand Central Parkway, and we could look from the top of one of the bins over a high fence across the street to where Jimmy Walker and Betty Compton used to meet. Poor Jimmy Walker. New York City perhaps deserved a mayor like Jimmy.

I was lucky one time to miss a subway train that crashed, and lots of people were killed and injured. Jimmy was on the spot in a few minutes, helping in the rescue work. Of course, after him came reform, Fiorello H. LaGuardia, and then, let us say, the new reform and the non-reformable Dwyer.

Then, there was James Michael Curley of Boston. Perhaps he took bribes, but he died poor; he gave it all away. When he was running for governor in 1934, I had a tire cover on my car, 1934 V8 Ford, with Curley's picture on it. I was superintendent for Perini on the Cambridge/Arlington section of the new Concord Turnpike.

The General Superintendent chided me slightly on Curley's picture. "Don't you know Perini is backing General Cole for the nomination?"

I said I knew that, that Cole wouldn't make it, and that my father had been ill and unable to work for a while, so I expected Curley to get him a job. Curley won, and I remember, at a road builder's dinner just before the election, he said, "If I am elected, I shall build sidewalks on all the state roads." And he did.

There was some talk because he bought a lot of short granite pieces from Fletcher Granite quarries to do this. In later years, one of his Republican critics had to praise the project because, besides protecting pedestrians, it also stopped cigarette butts from blowing off the highway and causing forest fires.

My old English professor at MIT, Tubby Rogers, used to say that Curley cost the city more money than some of the other mayors. He gave the city an extra $1 million of administration and was worth all it costs. This professor was the same one who told the seniors in 1929, "Be a snob. Marry the boss's daughter. Don't try to marry his secretary."

Of course, although Curley hated Perini, who had opposed him by backing Cole for governor (he used to call him "that guinea"), Perini teamed up Joe Maney, who was an old Irish contractor who was very friendly with Curley and formed the two companies. Perini and Maney built the main dam at Quabbin. All three, Curley, Maney and Perini, played ball. I was not on the job, though I had been in the earlier work in the area.

In 1938, when I worked on the Lincoln Tunnel job at Union City, New Jersey, I was interviewed and asked questions by some IRS people, including Elmer Irey, the famous Treasury investigator. These people showed me a list of bills, mostly from trucking contractors, and asked me if I had approved any of them.

I told them I had not, and it would be easy for a superintendent on a big job to approve large bills for such hired trucks without remembering the specific names. Also, as a matter of fact, I recognized one name, but this trucker had not worked for me, so I did not know what the story was. After the meeting with Irey, I went to a pay station and called Louis Perini and told him what had happened. He said it has nothing to do with me.

Later, I found out that he and Maney had written some checks for some nonexistent trucking company, and Eddie Cavaugh, Louis's brother-in-law, endorsed them and then passed the money along. Of course, they carried the bills as a trucking expense so as not to pay tax on the money. In the long run, Maney and Perini were convicted, not for bribery, but for income tax evasion, and it cost them a few years later, when the U.S. entered WWII and defense projects started up. They were banned for a time from bidding these projects.

By that time, I already knew that paying bribes was a losing game. If you have to pay off to get the work, and that is sometimes true, you are faced with several choices. One, you pay the bribe, do the work well, and pay tax on the money as income. This way will probably cost you money. The second choice is to cheat on the quality of the work and so absorb the loss and evade the income tax. This may get you into prosecution, fines, and jail.

The basic rule to me has always been that anyone you can buy for a $5 lunch can't be very useful to you or to himself or to anybody. All I can do now is tell you that old story of the commissioner of public works in a city I knew.

His secretary was concerned when she heard a loud altercation in his office and saw her boss throw a man out of the door. The commissioner was very upset.

He told his secretary that this man had tried to bribe him. She said people had been trying to bribe him for twenty-five years.

He said, "Yes, but this man was getting dangerously close to my price."

I suppose no one of us should be smug about our personal virtue, if we really know what our price would be.

Pipelining to me is a real sense of movement. All construction, any construction, can be built with movement. But consider pipeline movement: there is not real static portions.

The front fence gang goes out with trucks, tools, and fencing materials. They cut the first fence, put in a gate with an opening wide enough for all equipment and men following the paths. Then, the gang moves to fence number two and so forth, until the gates are all set. Before they finish the section, the right-of-way crew, with their bulldozers and drag lines, are moving, leveling side hill cuts and easing grades to minimize overbends and sidebends.

Then, the stringing crew comes in, and they start right away stringing pipe. The survey engineers stake out for the bending crew. Welders on the "firing line" weld the pipe sections in the longer stalks, and the bending crew with the ditching machines are right with them. Next, the tie-in of the pipe-laying crew. They make a continuous line. Pipe cleaning and priming crew are right behind.

Then, the paint or dope crew come along. The lowering-in gang follows and puts the pipe in the ground. Then, the backfillers backfill it, then the cleanup crew who clean up and regrade, and finally, the back fence crew replaces all the fences. When the back fence crew starts, some of the other crews are already moved on to the next job.

The circus crews moved in a similar way, except for the pause of the show. But even while the circus show is on, there are parts of the new crew out moving along. The Panzers of the Nazis studied the circuses; they could have studied the pipeliners, too, and perhaps did. A pipeliner does not sleep long in an area. The office trailers are usually moved every two weeks to a point halfway in front and later in back, as the crew lays perhaps a mile a day.

The workmen do not find many lasting relationships along the way. Some, like Big Hickman, looked for a clean married woman, who could not, for fear of the talk of the town, cheat on her husband with a local yokel, but the moving of the pipeliners makes him like a knight errant. The relationships are brief and sometimes tender.

Romance Scott, a welder, used to fall in love every two weeks. But it always passed with the motion of the spread. Baldy Hoyt once looked for a room in Ohio, and the man of the house said, "My wife is out buying groceries. She takes care of the renting. By the way, what do you do?"

And Baldy says, "I'm spread inspector on this pipeline."

"Out!" said the man. "I don't want no pipeliners here. The last one here wasn't here two nights before he tried to screw my wife and did."

One of the things said about pipeliners is that they are victims of the so-called "pipeline rumors." I was trying to keep some welders from leaving before a job was finished. They said there was a job in Louisiana that would last all winter and were going to set in there.

I said, "Haven't you read the papers? That part of Louisiana is all under water. It will be a long time before they can lay pipe."

"Don't care. We're going anyway."

The classic story about pipeline rumors concerns a pipeliner who died. As he approached the Pearly Gates, St. Peter came out and said, "Sorry, we're all filled up. No place for you to get in."

The pipeliner sat outside the Gates and looked longingly in. Finally, he got an idea. He thought, *I'll start a pipeline rumor and get some room.*

He saw a pipeliner and called him over, saying, "You know there is going to be a big-inch pipeline that's going to be built down in Hell." The pipeliner went off and soon collected a group of fellow pipeliners, who told St. Peter they were going to see if they could set in that big pipeline down in Hell.

Suddenly, there was room, and St. Peter let the pipeliner in. But he was bored and unhappy and couldn't find any pipeliners to pop off with and talk about pipelining. So, one day, he went the Gates.

St. Peter said, "Where are you going?"

"I'm going down to Hell. They just might be going to build that pipeline there."

Like tunnel stiffs, the pipeliner has nicknames for many: Angle Iron Slim, Sucker Rod Pete, Romance Scott, Two Day Whitey. Two Day Whitey was a fine pipeman and also a gambler. He had always set in at a job, but only stayed two days: the day before payday, then he'd win lots of money; and after payday, when he'd leave for another spread. Somehow, he got sort of attached to me and stayed three or four weeks on a job.

Of course, besides the cross-country spreads, there are the specialties: river crossings, yard piping, compressor stations, and pumping stations. The best river-crossing superintendent I ever had was Angelo Tavernaro, an Austrian from Tyrol. When he was doing a river crossing, well points, sheet piling, or dredging, when he could see and keep track of everything, he did great. But put him on a cross-country spread where he had to take in a full day of driving hard to see each operation, he couldn't cut it. Road and railroad crossings and boring crews are specialized sorts of work, too.

Mr. Foreman met me at the air field at Olathe. We had several jobs, and I went to work traveling as before, from job to job as general superintendent. We had a water plant and pipeline in

Missouri, a rehabilitation job on a products line in Chicago, and two gas compressor stations in Kansas with Northern Natural.

Then, I tried to get separated at Leavenworth in Kansas, but the base commander at Geiger called me up and said I must come back. All he wanted to do was con me into staying in the armed services and start training a bunch of recruits. But I had lots of points and finally got separated and went back to work as full civilian.

I had been brought up on the beach, and the lure of the Boston area seemed larger to me, although all the time I was in the Army, I was thinking about going back to the pipeline again. Then, I was in Excelsior Springs, Missouri at Christmas time. The roads were all ice and the airplanes not flying, and I came down with malaria. I couldn't get home to see my wife and kids. This didn't help.

Anyway, I went back to Kansas City and talked to Charlie Foreman. He wished I would think it over. He had no son; he thought I could do okay.

I went back to the Boston area, and after some more thinking, I bought a big house with fifteen acres and started raising more kids and doing a bit of consulting for contractors and some subcontracts on bridge repairs. Some of it was a highway in Fall River, a pier in New Bedford, a dam in New Hampshire, an airport in Fall River, and bridges and roads in the Berkshires.

♦ ♦ ♦

*Cape Cod Canal, 1936*

*Cape Cod Canal Widening, 1936*

Tuscarora Tunnel

Tuscarora Tunnel, JJS

*Lincoln Tunnel, 1938*

*Lincoln Tunnel*

*Pittsburg Dam-1*

*Pittsburg Dam-2*

*Pittsburg Dam Diversion*

*Bridge at Neponset Street, Worcester*

*Worcester Pike 1931*

*Worcester-Boston Turnpike, 1931*

*Worcester paving*

*Barber's Crossing, Worcester*

♦ ♦ ♦

# CHAPTER 3

## Army

When I went into the Army, I was in Ohio. My company had several major river crossings along the War Emergency Pipeline (big inch) and 400 miles of main line. A procurement officer told me, if I would apply, they would commission me a major, send me to a pipeline outfit, and get me overseas within ninety days. I believed him and volunteered.

Anyway, they did not make me a major. They said I was too young to get a direct commission at that time so they made me a first lieutenant. Then they said there were no assignments at the pipeline companies; they were full. I remember talking to a lieutenant colonel who was doing something about pipelines that he called Dresser couplings for Dresden couplings, so I imagined they filled the pipeline units with people who'd never seen pipelines.

The Army sent me to Fort Belvoir for what they call basic officer's course. We were in rank from second lieutenant to major and of course various specialists. They tried to cram the basic training for engineer soldiers or infantry plus the officer candidate school for ninety-day wonders into sixty days.

The physical part was not bad: I was in good shape, I was a good shot, plus my experience in explosives, camp construction, map reading, bridge and trestle construction, and aerial photos kept me on top of the courses. I didn't have to crack any books, and I could visit my wife every night, except when I had duty or sometimes a night problem.

Then, the Commandant of the school sent for me. He ordered Captain Decker to come, too. The only thing we had in common was we had wives in Washington. I was scholastically on top of the class, and Decker was at the foot.

Decker had been a hotshot construction man and superintendent of the building of North Station in Boston. He had been an enlisted man in World War I. The Commandant had no worry or concern about my or Decker's absence from camp. But he talked to me about an assignment in the States at someplace like Caterpillar Tractor.

I told him as a general superintendent and manager of a pipeline company, I had guys working for me at Caterpillar, Buckeye Ditcher, Lincoln Electric, and so forth, and I didn't come into the Army to be an expeditor. Besides, living and working in the States was more expensive. Overseas, you could send all your money home. I told him I couldn't get into pipelines.

He said, "Yes, they are full."

I said they only gave a lieutenant when they promised me a major.

He said, "Well, you are too young, but if you take this assignment, you will be made captain soon."

I said the only thing left the Army had promised me was that I would be overseas in ninety days.

He said, "You want that?"

And I said, "Yes." I was on a transport in eighty-nine days.

The colonel told Captain Decker that he was not doing well in the class at the school. Decker said he knew that and that, if he had wanted to go to school, he would have done it when he was young. The Army released him, and he went back to civilian life.

At the class, we had several fellow students who seemed okay. Klinefelter and Kelly were my best companions. Otherwise, the two months were pretty much a bore. The group was split up after course completion, and a few went to Ft. Leonard Wood in Missouri: Klinefelter, Smith, Marcantonio, and a few others whose names I don't remember.

I was assigned to a room with Smith. Smith was a Pennsylvanian Dutchman and, for a construction contractor, an odd sort. I was finishing up a plan and schedule that I had worked out for Foreman for preventative maintenance for all the company's equipment. The Army was also working on preventative maintenance schedules. Smith did not know much about it, but he was impressed, and he told the colonel about it.

The colonel said I should teach a course in preventative maintenance to all the officers in the group. The job fell on Smith, who was a captain, because the general thought that the course should be taught by a field-grade officer. However, Smith got the job.

He wailed to me, "See what you got me into?" He did not know where to start. So I got a list of training films and so forth from the film library and set up a list for him, so his course ended up as a movie show each evening.

The general decided to visit one night and found the room in darkness. All the officers were asleep, with Smith also asleep alongside the projector, which was running happily along, showing film strips on tractor maintenance. (I was not present.) The course was considered wasted on any officer less than a captain. Anyway, I believe the general cancelled the course.

The general was an advocate of a theory that the officers should belong to the Officers Club. We didn't like this, because the food was lousy and expensive. I was in a training company and could get meals with the company very cheaply, and it was good food. The club could not buy things through the Army commissary and had to buy at a civilian market. The low point was when we got what looked like chicken cutlets in a sauce that turned out to be just chicken necks, which was all the poor club officer could get in St. Louis for the money he could pay.

I still could not get into a pipeline outfit. Somehow, I got to see a colonel in Washington who was in charge of assignments for these duties. He didn't know any more about pipelines than a "pig knows about Sunday." He told the story of an unloading dock that he had designed, long and narrow so no one could drive a Jeep out to the end and would have to walk. I suppose it might have been good for people to get exercise for walking, but it might have slowed down the operation.

While I was at Fort Leonard Wood, my friend Mark Klinefelter and I had training companies. I was teaching demolition, mines, and booby traps to most of the new recruits, while Mark was doing the same for rigging, etc. It was boring.

Then, personnel circulated the division and wanted officers of a certain age group, rank, education, and experience to go to a school the Army was holding at Ft. Screven in Georgia, for training in underwater demolition and ship salvage. We heard about this, heard not only that it was an interesting course but at a warm-water beach. This contrasted with the woods of Missouri, where there was nothing but brush and mud and snow.

Mark and I volunteered. He had been a master diver, and I had some experience in compressed air in tunnels, plus a couple dives in hardhat gear. We did not get to this school. They sent a captain, who was forty-two and a piledriver man with no education. Mark was selected to be a radio announcer at the rifle range: "ready on the left, ready on the right, ready on the firing line."

I came into my orderly room one evening. The first sergeant looked at me with a curious smile and said, "Sir, you have orders on your desk. You're going to school!"

The orders were for Cooks and Bakers School. and I spent a couple of weeks learning about dehydrated foods. It was a good course; I learned some valuable things about the use of dehydrated foods. The diploma I got qualified me to be an "instructor in dehydrated foods." I was the only engineering officer at the course, among the others, who were female hospital dieticians, mess sergeants, and a couple of quartermaster officers.

But before the ninety days were up, we were shipped out to Camp Reynolds in Pennsylvania. This was a staging area. There were no regular duties except to get shots and so forth. Someone had set up what they called engineer tasks. These were projects assigned, like clearing brush and grading roads. Each officer was given a few squads of men to do these things.

I did not make the camp training officers happy, because I thought it was on the level and organized my group as if I had been on regular civilian construction job. Therefore, I got all the work done, and they had to find new projects. They told me that we were not supposed to finish. These jobs were only to mess around, so the next group of officers and enlisted men would have something to do.

This did not last long. Soon, we went to Hampton Roads and shipped out to North Africa. We went out as an officer cadre to take over the units of the 7th Cavalry, Custer's old outfit (all Black, except for the officers), to replace the officers and make the cavalry into engineer soldiers, construction units, and so forth.

This did not take place. During the three weeks that we were in transit, the idea changed. The Army decided to keep the same officers in these outfits, and we had no place to go but replacement depots. Here, I decided to go to a hospital for a minor operation for hydrocele that had been caused by a training accident at Belvoir. I did not wish to miss the boat, so I had a civilian doctor drain it for me.

When we were loading the ship, the Army doctor looked at me and asked, "You're not ruptured, are you?"

And I said, "No," and off I went.

While I was in the hospital, I learned that Klinefelter and some of the others were assigned as first lieutenants to command platoons of firefighters. However, by a stroke of luck, I found that my old friend, Jim Lambert, who had been shovel boss for Perini in the early thirties and then left and founded his own company, had gotten into the Army and was lieutenant colonel in a construction outfit that recruited from the Italian engineer troops: Construction Battalion 7077(I), I for Italian in parenthesis.

Most of the Italians had been POWs, but they had been sworn into the U.S. Army, and the U.S. was paying them. Jim got me transferred to his headquarters, and we did work on railroads, bridges, docks, mine clearing, quarries, and so forth. The Italian enlisted men and noncoms were good workers. The officers, on the other hand, were useless and, with few exceptions, were political appointees. Before we left for Italy, there were short periods of doing various construction jobs. We look a few men to Oejda, Morocco to pick up and clean up for the Fifth Army, which was leaving for Italy.

The British and South Africans operated an airfield there in Oran, Algeria at LaSenia, Spain. We had two projects connected with this airfield that were interesting. One was a U.S. artillery spotting aircraft that had landed in the Sebhka D'Oran, a swamp and salt marsh mud flat adjoining the air field. This plane had landed short of the runway and was stuck in the slippery mud.

Someone had called the engineers, and a D6 tractor had been dispatched to pull the airplane out. The bulldozer bogged down and got stuck in the mud. Then, we got the job. Captain Kemp, a contractor from Virginia, got about 400 POWs and lots of rope, and in Volga boatman style, we pulled the airplane and then the tractor out.

Later, the British had dropped a Lancaster bomber into a race track infield near Sidi Bel Abbes. This town was headquarters of the French Foreign Legion at the time. I took about twelve of the Italians and a US sergeant down to get it out.

The aircraft was in the middle of the elliptical infield. The track was only about a half mile long. To get it strip graded long enough to get the bomber off, we had to excavate part of the track at one end and fill in the other end, plus also cut down a few 500-year-old olive trees. The owner was a Frenchman who also ran a few Turkish-type bathhouses in the town of Sidi Bel Abbes.

We got our men and equipment there: bulldozers, patrol graders, roller, and so forth. The RAF had a crew of twelve aircraft mechanics who were removing all the armor from the airplane, to lighten the airplane and tune up the engine. The pilot was not present. He was an Australian lieutenant, and the airplane was to be flown out by the group captain from Blida headquarters. They wouldn't trust a Colonial to do it.

Well, we got the strip in shape, and I left the area with twelve of my people living under one wing and twelve RAF under the other wing, all happy because some of the native girls were bringing them wine and other good things. When the group captain finally came over, the aircraft got off safely, and I went down to check that fence and track were put back in shape. The old olive trees probably cost Uncle Sam plenty.

We sailed to Italy with two companies of Italian troops, and I had a U.S. lieutenant and four U.S. sergeants with me. We later picked up enough Italians for a full battalion. We finally got to a small town near Caserta called Caizzo. The troops were quartered in an old monastery that had been pretty well bombed out. The whole town had neither running water nor electricity.

My lieutenant and four sergeants and I took over a villa complete with patio and lemon trees. There was an Italian lieutenant colonel, medical, who was assigned to the battalion. He was a very well-educated man, far more than the line officer was, and they resented him. I took him into our villa, and in the evenings, he was a very pleasant companion.

My lieutenant was a very diligent Jew and wrote long letters to his family in the States. I don't know who censored them; they were always written in Hebrew. He was a very able officer and knew army routine very well. He was not in the least knowledgeable about engineer work; he was a chemist at the water treatment plant in the US. Because water treatment was an engineer function, they commissioned him as an engineer.

We had to build a pipeline bridge over the river, and when I told him, he said, "Oh, that's easy to build a bridge." He had gone through an exercise at the engineer school at Belvoir, where pre-cut timber was used to build a trestle across a dry ditch. He found out that real bridges over real rivers were different.

He also had been given a course on bulldozers. The training film (I saw it later) showed D4s in close-ups and D8s in a distance, pushing down big trees. He thought that the eight D4s we had in our battalion must be great.

He used to listen to my theory of getting things done as had been described many times by C.S. Foreman: "Do something. When you are faced with a choice and have two alternatives, you must pick one and do it. You have a fifty percent chance of being right. If you are wrong, you can change. After a while, you will learn more and have a better average, but if you don't do anything, you are 100% wrong."

In the town of Caizzo, there was a nunnery. There were a lot of priests in town, too, even though the monastery had been abandoned. I met an American girl, a nun, who had been in Italy when the war broke out and gotten stuck, so she'd had to stay at the convent. Her superior was a large, plump Italian woman with few front teeth. She spoke no English, just a sort of fractured French that was harder for me to understand than her Italian.

The Germans were making some kind of push in the north. We only had side arms and carbines for the lieutenant and myself, and the sergeants had carbines for interior guard duty. We had to lift quite a few mines every now and then, but we were not combat troops.

Regardless of this, we had orders to allow no leaves for a period. We Americans, unless we got R&R (rest and recreation), could get no leave anyhow, but four or five of the Italian officers decided to take a truck and go to Rome. Earlier, when we were near Naples, one of the Italians asked leave for forty-eight hours. He went to Sicily and did not come back for ten days.

I had taken my .45 revolver, which I usually kept in the glove compartment of my weapons carrier (a ¾-ton I found better than a Jeep for getting around in the mountains). I ordered this Italian lieutenant to the rear seat and had my Italian corporal driver take us to the nearest stockade. This stockade had a few Italian POWs not yet released and a few Russians and Poles.

I told the camp commander this man had broken his oath of allegiance to the U.S. I considered him still an enemy and an escaped POW. I had summary court authority over the enlisted men, but for an officer, you needed a general court, and we had no such facilities. I could not break down the morale of my troops by letting an officer off for a violation that would have given an enlisted man thirty days in stockade at hard labor.

The stockade commander gave me a receipt for this lieutenant and I left him and never saw him again. This made an impression on the troops and officers alike.

Spending the night in Rome was the objective for these four characters, and they expected to come back the next morning. They had taken the worst truck in the battalion, an old orphaned Dodge for which there were no spare parts. The truck broke down, and they walked a few miles to get help. While they were gone, someone stole the vehicle. We looked for them and for the truck with no success. I went to Rome and reported to the MPs, but they never found it.

After a few days, I heard these men were back in town, probably to get food and shelter, but I would not move. Then, the little nun came to see me, along with her Mother Superior. They asked me if I knew the men were back. I told them I did and that, if I saw them, I would have to capture them. But if not, I was not going to hunt for them.

Both ladies made a plea for these poor officers. They said they just wanted to see their mothers. I said the men were lying—none of their mothers lived anywhere near Rome.

The old nun said that they feared, if they came back to me, I would strap on my pistol and take them to the stockade. I said I certainly would, and I hoped they would not come back, because I didn't want them, and I did not wish to feed or house them.

I asked the American girl why she was doing this. I told her that I knew nuns were supposed to be kind and make suffering less for all people, but why would they interfere in such a case?

The American nun said the bishop had ordered them to talk to me. I said it was sad that these officers were hiding behind the skirts of the nuns and that the bishop was not man enough to come and talk to me himself. If he was going to act this way, I would put him in the stockade. This was all in English, but the old Mother Superior broke out laughing and made motions depicting the bishop behind bars. She was grinning a mile wide; she evidently had no liking for the bishop.

We moved up to Livorno (Leghorn) by LCI, along the west coast of Italy. LCI are flat-bottomed boats called Landing Craft Infantry and can carry about one company of men. I rode in the lead vessel. The officers were all young ensigns, all four. The skipper offered me the lower bunk in his room. I guess he imagined he wouldn't want me to fall on top of him.

The trip was very rough. One of our Italian soldiers got such a bad case of seasickness, he had to be put in a hospital when we made port. On this trip, we were near Elba when I was awakened by a sailor who thought I was the skipper, because I was in the skipper's bunk.

"Sir, Mr. Colbert wishes to see you at the bridge." I thought my troops were in some sort of trouble, but when I got to the bridge, I found he did not want me anyway. He wanted the skipper to take the conn while he went to the head.

We moved into an area at Marina DiPisa, a beautiful spot, full of fine pine trees belonging to the king or that had belonged to the king. It was just inland from where a beach had been pounded by the U.S. Navy and only a few pillboxes remained; when I first saw the site, it was empty and damaged. There was a stock fence along the road.

I told the major showing me the site that, before we set up camp, we would have to sweep for mines.

He said, "Who would put mines among all those big trees?"

I looked along the fence and said, "Those could be famous last words, Major." Then, I crawled along the outside of the fence and reached the inside.

There, I cleaned a patch of grass that appeared a bit browner than the rest and uncovered a mine, one of the glass varieties. There were so many mines there that they gave us a company of Canadian engineers, sappers, to help out.

At that time, when you lifted a mine, a teller mine (*teller* means plate in German), you could check the three wells it had around the base. If there were trip wires or booby trap mechanisms, you could cut them and remove the mechanism. The main actuator or exploder was on top of the mine. This exploder would be disarmed by turning it and unscrewing it. Then, the mine was deactivated. It was merely eleven pounds of TNT and could not be exploded, except with a high-power blasting cap.

Inside the detonator, the actuator was a .45-caliber blank cartridge that would fire into the primer, making the whole eleven pounds go up. Once you took the detonators out, you piled the deactivated mines along the road. If you needed explosive for quarry work or blasting boulders, this was a good source.

However, we very shortly stopped lifting the mines this way because the Germans had a new pin inside the detonator so that, when the mechanism was turned, it would break the pin and fire the mine. After that, the only safe way was to lay a half-pound block of TNT alongside the mine or take a long rope, pull it out of the ground, and explode it. We often wondered how many engineer soldiers got blown to bits before this latest German invention was discovered.

The work in the Pisa area for our battalion was: 1) to construct a staging area to take the troops out at the end of the war, which at this time (late 1944) was assumed to be early 1945); 2) to operate rock quarries to make crushed stone for the continuing advance of the Fifth Army; and 3) to build a large hospital at an athletic stadium near the famous leaning tower. We had several other projects, as well: a pipeline bridge over the Arno and water line, among others.

I suppose, at this time, I could have been called a general super for a construction company in the States, except for the great difference in pay. My staging area project needed lots of lumber, and there was plenty in the depot north of Livorno. But, like all depots, no one wanted to let any out.

We had three U.S. Army sawmill companies getting out about 20,000 board feet a day each, and I requisitioned an Italian sawmill that did not turn out more than two thousand board feet a day. To meet the schedule for completion in March 1945, it was necessary to have 80,000 board feet per day, and we were only getting 62,000.

We were attached then to the 338th Engineer Battalion (construction), commanded by a Colonel Hahn. Hahn was a West Pointer, but I believe he had spent more time coaching football than performing as a Corps of Engineer officer. I continually had to requisition through his S4, his supply officer, for lumber and so forth. This officer would accept the zeroing of the requisition by the depot.

My complaints finally got to the colonel. He said, "I will give you D Company to help you out."

I said "Colonel, I don't need any more men. I need wood. Please, come out and see for yourself."

He said, "The more the merrier," still thinking that more men could help.

Anyway, he came out the next day, and all was going fine—saws were sawing and hammers were pounding. Then there came a lull. A trailer came in with a few boards, and the men grabbed them. The noise started, then the noise stopped again. However, the colonel got the idea. The next day, we got the papers we needed to get all the lumber we needed. (The irony was, a few months later, the depot burned down, along with all the lumber.)

The hospital at the stadium was designed to put operating rooms and X-ray rooms, wards, and so forth under the stands of the stadium. Some structures were laid out on the playing field. There was a gatehouse, a guard station, probably a ticket office that we planned to rehabilitate for security, and temporary offices. This building was full of squatters, all Italian civilians.

It would not do, I decided, for U.S. troops or U.S.-led Italian troops to dislocate these people, so I wrote a note to the local commandant of the Carabinieri, telling him of our problem and requesting him to send one (uno) carabinier to evict them. Carabinieri are more than police, sort of like Texas Rangers: one riot, one ranger. So uno carabinier arrived the next day, spoke firmly but quietly, and the problem was solved.

My driver, Micheletti, was a professional driver of trailer trucks in the mountains. Some of the U.S. officers used to rib me, because I had the best driver in the battalion as my chauffeur, but I told them I deserved it. This Vittorio had been assigned to the German Air Force (Luftwaffe) as a tail gunner. He had been shot down and taken prisoner. He had very rigid principles. He had contempt for all Italians from south of Verona, and he considered the Italian officers who had stolen the old Dodge truck as *disgradia*—disgraceful.

One rainy day, we were driving along a road packed with sheep. We had some kits of copper welding fittings in the back of the ¾-ton truck. Something bounced in and out of the back of the vehicle. We stopped. A sheepherder had taken one of the kits. I grabbed him by the front of his raincoat and shook him. The kit popped out. I made him give it to me.

Then, I looked back, and my driver had taken my .45 revolver from the glove compartment. He handed it to me. "Shoot the bastard, sir."

I did not shoot him, but the sheepherder looked petrified. Micheletti was very sullen for weeks—I had let him down.

He never learned any English; we always spoke Italian. Except, each morning, he would salute and say, "Good Morning, sir," in English. Quite a character.

The hospital had been designed by the engineering command, and we, of course, were not supposed to alter anything by a foot or an inch. With construction well underway, we had a visit from the medical people: doctors, colonels, and so forth. They were not happy with some of the layouts of the operating theatre, the scrub room and more.

One famous surgeon showed me how it should be set up. Scrub, hold your hands high, then open the swinging doors with elbows. The radiologist showed us how the lead sheeting should be placed in different areas, so it was economical yet safe.

Then, one of the medical people looked at the plan and said, "Where's the mess hall for ambulatory patients and enlisted hospital personnel?"

I pointed it out and asked what he thought that was. Well, he'd thought it was a parking lot, but he didn't know how to read a plan. I asked the chief surgical officer why he had his name on the plans, and he told me he couldn't read a drawing, either, and did not know what had been designed.

Anyway, I made the changes so the hospital would be operable. Then, one day, when the turnover of the facilities was due, a bunch of the Inspector General's people arrived. A warrant officer had a big roll of plans. I almost got court martialed for making the changes. The changes actually saved manhours and materials.

The IG Colonel, when I said the medics had needed it, told me, "The hell with them. They signed the drawings."

However, we escaped, and the hospital was accepted. And I didn't get court martialed.

About this time, the commanding officer of the Technical Supervision Regiment to which we belonged showed up in Leghorn. He had stayed in North Africa long after he had any business to be there. All his people were in Italy, but he had found a very nice French lady and was in no desire to leave. He was like a school boy in love. He told me he had never had such a great love and was trying to figure out how to get rid of his old Stateside wife.

Anyway, he sent word to have me see him in the morning. I had an excuse, so I sent Lt. Schwartz. Schwartz came back very subdued and unhappy.

The colonel brought up the treatment of the Italian officers many months before. He told Schwartz I had no right to do what I did. Schwartz had told me that long before this, but I had to do it anyway.

Schwartz had absorbed some of my philosophy of doing something and so forth, so he told the colonel, "I was faced with a problem. I had to make a decision. So, I did it."

The colonel chewed him up and down; he did not, however, tell him how it should have been done. Then, he chased him out and told him to have me come down and see him right away.

By the time I got there, the old man was calmed down. He talked about his French lady briefly, and then he complimented me (not on the subject of the Italian officers) on the reports he had received on the performance of our troops and their work. I never saw him again.

There was a big quarry near Lucca. We were told the Fifth Army needed 1,500 tons of crushed stone each day, and we should get it from this quarry. It had been operated by townspeople for years, and the commanding general in the area seemed to think we should operate it with the help of about 200 civilians. I said I did not need the civilians.

I was told that they had been running the quarry for years. Whereupon I said they had been running it wrong. Then the 338th wanted to give me a company of theirs to use. I had run quarries in Africa with a few of the Italian troops, and I insisted that all I would need was a half platoon of my Italians. Except that the 338th had a corporal who had been a silver miner in Nevada, and I could use him to help drive a coyote hole (a small tunnel near the bottom of the face) and make a T on it, then load it with powder and pull the whole face down.

In the interim, I could use six wagon drills to cut into the corners, to keep the flow of broken rock to the crushers. The depot had the crushers and conveyors and screens but could not find the wagon drills. We started the tunnel with a couple of handheld jack hammers and wheelbarrows. When we found the wagon drills (the Army called it a drill wagon; while civilians called them drifter drills, wagon-mounted), we couldn't get the number I wanted. I was told they were already allocated, though I never found out whom they were allocated to.

They had twelve, and I wanted six: four drilling, one moving, and one under the drill doctor's care. Drill doctor is the name hardrock tunnelers use to describe the mechanic who keeps the drills maintained. Anyway, they said I could only have three, because "they were allocated."

In the meantime, we had assembled the crushing and screening and stacking equipment. After I finally almost got court martialed, I got off the duty. Captain Kemp got another quarry going for our own hospital staging area work. The 338th assigned one of their companies to the job at the quarry.

After a few days, the company commander called me and told me they had the three wagon drills but couldn't find the lug shank hollow drill still needed. I told him to have the corporal who was driving the coyote hole go down to another depot, where the right steel was lying right next to the highway. He did not wish to ask an enlisted man's help, so he offered, and I accepted, the right to buy his monthly ration of whiskey. In our Italian outfit, we got no liquor ration.

Anyway, he called for me in his command car, and I showed him where the steel was then asked him why he could not find it, when it was so obvious, he had to almost run over it. He said he thought it was some kind of pipe.

His driver had to laugh, and I suppose the captain found some way to punish him. I saw the man later moving up with a combat unit.

After some time, the captain came to me again and said the corporal was almost ready to shoot, and he wished I would come down and check with him on how to load it. I went down but could do nothing. The company had set up crushers, stackers, etc. in such a way that the coyote hole shot would bury the crushing equipment with broken rock. Before they fired the coyote tunnel, all the plant equipment had to be moved. I don't know what they ever did.

That battalion had a poor bunch of officers. We were supposed to use their messes and quarters, because we did not have enough men to qualify for a separate mess. After one or two meals with them, we joined in with the Italians and drew Italian rations. We didn't get such things as butter or turkey, but we made do.

At Christmas time, I sent over Sgt. Daum, who had a plate in his head from Salerno but was a good scrounger. He went to see the major at the quartermaster depot and found out they needed some soakage pits for sewage. So, we sent down a few men, plus a compressor, a couple of jackhammers, and a blaster to blast the pits. I can tell you, that Christmas we had more turkey and fixings than we could eat.

Speaking of depots and supplies, as mentioned, we had a small quarry operation. We had a crusher that used an engine the size of a D6, and it was broken down. I went down to the depot to see about getting a replacement. The people there said, "You don't have that on your table of equipment authorized, but if you bring it down, we'll look at it and see if it is worth fixing."

Well Capt. Kemp and I saw that would be hopeless, so we looked at it. When we thought about it and the quarry at the same time, a light dawned on each of us. As I said before, we had D4s in our table of equipment, but they were not much use. We had scrounged up a couple of D7s for bulldozer work. Then, we had a great idea: we'd put a D4 engine in the crusher, wedge back the governor with crushed rock until the D4 engine crapped out, and then put it back in the D4, replacing it with another engine from one of the other D4s and hauling the old D4 with the blown engine back to the depot, where we'd exchange it for a new D4. If the war had lasted much longer, there wouldn't have been any D4s operable in the whole theatre.

About this time, my limited duty status ended, and I was transferred to the aviation engineers, 815th Battalion, which was in the process of pulling out and going to the USA or Okinawa. I only had a brief time with them before we were moved to one of the 15th Army Air Corps bases near Foggie at Manduria, in southeastern Italy.

The runways had been in use for a couple of years and were in good shape, with a tufa surface. We took over maintenance and drainage of several air fields. Shortly after, the colonel told us we could get some Bunker "C" oil, which has asphalt in it, in the Navy stores in Taranto and apply a coating to the runway. I figured we would need about 84,000 gallons, plus or minus.

We got the material in our tanker trucks and sprayed the runway. It came out fine. Two days later, our colonel came over and, after looking at it, asked me how much we had used.

I told him we had used 82,000 gallons. He took out his notebook and said, "Didn't you figure 84,000?"

And I said that we'd done it with the slightly lesser amount. He ordered me to go back and use 2,000 gallons more, which we did. It just fogged the area, and that was it. The Navy captain at Taranto was furious.

Then, Pres. Roosevelt died. We all stood in formation, and most of us were sorry to hear the old man was gone. We didn't go to Okinawa. Instead, the whole battalion was shipped to the US for forty days of rest and recuperation. Then, on to Geiger Field in Washington, to regroup and get ready for the invasion of Japan. With the 815th, I met Glen Maxon, and that was a success.

We were on the first ship back from Italy. After forty days of rest and recuperation, we started taking in replacements (about fifty percent) and went on to extensive training. Then in August, Harry Truman dropped the bomb. But when it came, we were still at Geiger. I asked for leave and received permission. I had a lot of leave time coming, and I went back to work for C.S. Foreman. I hitched a ride in a Navy R4D, which is the Air Force C47 and a civilian DC3 and which the British called Dakota.

♦ ♦ ♦

# CHAPTER 4

## Greenland

*Editor's Note:* The following is a listing of some of the Greenland projects from Metcalf & Eddy's (M&E's) brochure that are referred to in the narrative:

**Thule Air Base:** planning, design and services during construction of major airfield; communications facilities, including ionospheric scatter, telephone system, Nav-Aids; dock, utilities, and roads; fueling facilities; hangars, shops, operation and administration facilities; barracks, mess halls; warehouses, gymnasium; chapel; Nike-Hercules batteries, AC&W Station; ammunition storage. All these facilities had to be built within a four-month construction season. Also performed design and services during construction of Sondrestrom and Narsarssuak Air Bases, and Simiutak Navigation & Communication Facility.

**BMEWS* (Ballistic Missile Early Warning System) stations:** these large structures involved unusual foundation and weather-resistance problems. Each of the four antennas is the size of a football field set on edge. The 37,500 kva electrical substation for BMEWS must be able to operate at 40°F and in 150-mile gales.

*BMEWS was a network of long-range radar station installations built during the Cold War to provide advance warning of incoming Soviet ballistic missiles.

**Eastern Extension**: DEW Line. Siting, design, and services during construction of four Early Warning Radar Stations, each with tropospheric scatter facilities, employing 120-ft diameter "dish" antennas, with independent power sources and support facilities; one of two land stations has airstrip, other has helicopter pad. At each of two icecap stations, entire installation is a composite structure mounted on legs founded in the icecap, so as to permit built-in jacks to raise the station from year to year in order to maintain its position above the ever-changing snow surface (these were called Dye sites).

**Camp Century**: Design of appurtenances to a semi-portable nuclear power plant on the icecap, including enclosures for the spent-fuel container, hot-waste tank, feedwater equipment, decontamination facilities, laboratory for control of concentrations of contaminants, and design of a dissipation system for the removal of excess heat generated and its discharge to areas where it will not affect the structural snow in which the facilities are erected. Also performed the feasibility study for this construction with comparative cost studies of the heretofore unexplored problems of founding a nuclear plant on an icecap.

**DEW Drop:** Siting and design; services during construction for long-range tropospheric scatter link which has two stations, two 120-foot "dishes," power and equipment building, housing, and messing; fueling facilities, utilities, and roads.

*Editor's Note:* This memoir reflects terminology and usage current during John J. Scheuren's work in Greenland in the 1950s and 1960s. In some instances, references to Inuit people may not conform to contemporary usage. Today, *Inuk* refers to an individual person, while *Inuit* refers collectively to the people. Because these distinctions were not widely recognized in English-language engineering, military, and construction communities at the time, certain historical usages have been retained to preserve the authenticity of the author's recollections. Readers will find throughout these pages that Scheuren held the Inuit people in the highest regard and wrote of them with admiration, respect, and gratitude. No disrespect is intended by any historical terminology appearing in this memoir.

In early 1951, Andy Paton, with whom I had kept in touch, called me up. He'd left the contracting firm he had been with after the war and gone with Metcalf & Eddy (M&E), who had gotten in on a big job in Thule Air Base, Greenland. He wanted me to go up there. However, my wife was pregnant at the time, and it turned out to be twins. I did not wish to leave.

But by the beginning of 1952, the twins had been born (in mid-November, 1951), so I signed up and went to Thule. I was in the Arctic or Antarctic for about ten years, 1952 to 1961, and only home for three-and-a-half months during that time.

The runway at Thule was graded in 1951, but we had to pave it in 1952 and, in 1953 complete it with taxiways and hardstands, plus finish the hangars and POL (petroleum, oil and lubricants) facilities.

In 1951, we had developed a small lake about midway on the runway, where we set up three seawater distillation plants. The lake, named Eddy after Mr. Eddy, was small and could not be increased in size. The distillation plants were inefficient and never did meet the specifications.

We had specified certain manufacturers. However, a recently formed company called Mechanical and Electrical Inc. from New Orleans entered the picture, and although we never could approve their product, they got into the halls of Congress. Senator Long of Louisiana made a few impassioned speeches for his local constituents, until the Army finally bought them. They never got up to the design capacity and were full of problems.

Carryover of slugs of water damaged the shafts of the Roots compressor (the method was vapor compression) and several types of metal parts set up galvanic action corrosion that kept shutting things down. Between Lake Eddy and desalination, we were struggling to get water for the base.

I engaged myself with Jim Otis, one of M&E's best water people, to see if we could develop a new water source. We found an area we later called Crescent Lake because of the shape of the basin. I measured the runoff into several brooks. There is very little snowfall at Thule and no rain at all, but we found, if we built a dam, even if we got only the equivalent of half an inch of the melt water from the snow to impound, we could have a successful water source.

We built a dam; then, we had to wait until the core froze so as to get an impervious seal. We also built a bypass trench from the river that only ran a couple of months and was heavily loaded with silt from the glacial runoff across the moraine, which was mostly clay and rock flour. Because of panic pressure from the Corps, the contractor diverted the river into the reservoir, and of course, the water was very muddy the first year. After that, we never used the diversion trench and got plenty of clean water from the snowfall and the spring melt each year.

Andy Paton had left Bruce Dunbar in charge of the site. I was not resident at Thule at the time; I was engaged in Project Charlie Price. This project was operated by Col. Charlie Price of the Air Force and some electronic communications people, Major DeWald, and others.

We sited AC&W stations (aircraft control and warning systems). The first was at P Mountain, a hill a few miles out of Thule, near the ice cap. Others were to be on the Greenland ice cap. One at about 5,000 feet elevation, near Etan, and one 200 miles northeast of Thule at 7,000 feet.

While I was flying around in ski planes, checking the distillation equipment, I was not present when the contractor asked the colonel in charge, "Schaefer," if it would be okay to divert the whole river through Crescent Lake and then let it dump into the next fjord. This would keep the water away from his operation downstream. The colonel, who came from California, and our man, Dunbar, were both completely ignorant about the problems of water treatment and filling up reservoirs.

One day, when I arrived back from a trip to the ice cap, Dunbar told me that he had agreed with Schaefer and the contractor, and the diversion would take place the next day. I told Dunbar that Mr. Eddy, who was in Boston then, would kill.

He went to the colonel and I raised hell. Ultimately, he refrained from this foolishness. The colonel had never liked me, and after this, still less. Of course, it was mutual. The Corps of Engineers and contractors had put no thought to the kind of people they needed and had sent an Area Engineer who had never been exposed to cold, snow, or frost and a contractor superintendent who was equally ignorant.

The road to P Mountain was laid out by a fine road builder, Lou Litchfield, who had left his job as director of Maine Highway Department to come to work with M&E. He, of course, knew the principles of building in frost and snow. But Gammill, the contractor's superintendent, who had never worked in snow and frost, showed the colonel where Litchfield had laid out the road on top of the ridges and with no side hill cuts, stating if he, Gammill, was allowed to cut off along the side of the mountain, to side cut the material, much money would be saved.

The colonel, in his ignorance and much to the disgust of Lou Litchfield, concurred and then directed the building of the Gammill section of the road, as it was called (especially during the weeks when it was closed because of land slips and snow drifts) by the colonel who followed Schaefer. The stakes, which had been set by Lou Litchfield, could be seen from the road, free and clear of snow, while for years, and probably still today, the big snow removal process goes on.

During the flap about the distillation units, which happened when I had gone to Boston for a meeting, one of the Caterpillar diesel engines driving a generator for the distillation plant had "run away" and gone to pieces. This can happen to a diesel engine if there is excess fuel and the governor is not working correctly. The engine, if there is plenty of air, will run away and tear itself apart.

I rode back to Thule on an airplane with a tech rep from Caterpillar. We discussed it on the way, and we thought it might have been caused by a defect in the governor. When we got there, the Caterpillar man picked over the pieces and reassembled the engine. The fault was found to be in the operation of the Woodard governor. The governor had a cover plate that was bolted down by four screws. One of the screws was shorter than the others, so it would not lock down the rack gear and keep it from oscillating.

Probably during a regular maintenance check, the operator, who had not been properly instructed (because the duty in Thule was one year, and there was no overlap, so the arriving operator and the departing operator might arrive and depart in the same aircraft) had not placed the short screw correctly. The rack was in the locked position, and the fuel continued to be injected, so the engine ran away.

I don't know if the makers of the governor had ever changed the screw set up. But someone figured they should put dampers in the air intake so that, if the engine did approach runaway speed, the air would be shut out and the machine could not fire.

Not very long afterward, it was shown that, with all of our technology, people can screw up. The oil filters at the intakes, as they are in a car engine, require changing of both filter and oil. One operator of one engine tried to add oil without shutting down the engine, and he managed to spill the oil into the intake. The engine started to overspeed, the dampers closed, but the engine, needing air, reached its suction down into the crank case, and a regular implosion instead of an explosion took place. The engine was ruined. In this case, there was no failsafe for sure.

We shall never catch up the benefits of our technology until we have become successful with people and proper training in maintenance. We had problems with the proper operation and maintenance of the power and heating plants because of poor maintenance. Both of the generators and boilers were a continuing problem because of poor attention to the treatment of makeup water.

On the outlying sites, we had no makeup water but only water with Prestone for cooling at the diesel generator sites. Our main grounds were tied into the undersea pipelines and to a massive amount of iron and steel set in the deep waters of the fjord. But on the outlying powerhouses, the groundwater was poor, and the coolant (because of stray currents) got a very low pH. This acid caused corrosion. We found out we had to check all the engine bases and connections for tightness and poor contacts, plus check the pH of the coolant (so-called permanent antifreeze can be temporary with chemical storage). However, by carefully checking on pH and cutting down on corrosion due to poor contact, we succeeded in holding down the damage.

Another angle in maintenance emerged. All construction had to be based on protecting the permafrost. When it is frozen, it is a good foundation. Unfrozen, though, it melts literally into mud and water. Therefore, you use timber piles or pad or both. Construction of low buildings like an office or barracks had to place a few feet of sand and gravel called a pad, non-frost susceptible (NFS) material on the ground, so that, during the summer, the thaw would be contained in the NFS and the underlying permafrost would be protected.

In the case of heavy structures like hangars, we excavated by drilling and blasting to a depth that would not be approached by the thaw. Then, piles were set for a foundation for the hangar. These were backfilled with wet slurry of sand and allowed to freeze in. The floors of the hangars were supported by a deep pad of NFS. In this pad, we placed pipes running from upwind to downwind, so that, in the winter, this cold air would circulate and freeze the material of the pads, not the permafrost. In the summer, the valves were closed and the warm air could not enter. This frozen block of NFS would not thaw down into the permafrost and cause settlements.

This worked for the first year, and then we found the pipes were getting filled with groundwater that would freeze and block out the cold and winter air from circulating. The next summer, the freezing was not much deeper than the pipes, and we had settlement. This was a two-headed problem.

Because of shipping space, the pipes were made in half and bolted and nested together. Therefore, if any groundwater came in, it would leak into the pipes because the connections were not watertight. Before the construction, there was no groundwater movement. However, once the pavement was placed, the black surface of the asphaltic concrete would soak up the heat, and each taxiway and hard stand became a small stream. This let groundwater into the gravel under the hangars that would get into the pipes, and then, when it froze, it blocked the cold air.

We found that painting the blacktop surface with white reflective paint was helpful but not really enough. We finally recommended threading solid welded pipe inside the bolted corrugated sections to allow the cold air to circulate and get the proper freeze back in.

We saw many people come and go at Thule. From 1951 to 1961, it was pretty much a male operation. When we came there were only two foreign women: Alice Crowell, the wife of the weather service man, Jack Crowell; and the wife of a Danish liaison officer, Commander Eric Staermose. Each family had a small house on the base.

The Crowells were childless, but in 1952, the Danish family had a little boy about three years old. Even that early on, we knew sled dogs were dangerous. Little Eric had an Inuit dog, a little puppy who had no association with the sled dogs at all. One day at play, Eric fell down, and the puppy, now half grown, started gnawing at the base of his neck. He survived but I suppose still has the scars on his neck.

The Inuit village Thule was named and established by the great Danish Inuit explorer, Knud Rasmussen, as a trading post. Old Ootah, who had been with Peary at the Pole, still lived there, a fine old fellow.

I had Robert E. Peary, Jr. working for us as a structural engineer. An excellent engineer but not the leader and manager his father was. He and I visited the village with a photographer one time. Peary's Greenlander son was alive then; he lived a few miles away from Thule. I do not think that he and young Bob ever met, although young Bob had been with MacMillan on the *Bowdoin* on several cruises. We sent Bob Peary as a manager and engineer to several sites. (I shall spend a few paragraphs later in a separate tale of the Inuit and our experience with them.)

Our people above a certain rank could be members of the Officer's Club. Andy Paton and I were charter members. It started out in a small way in a couple of BOQ's (Bachelor Officer's Quarters). There were, of course, no quarters for married officers. The designers did not see fit to specify any refrigerators or freezers, so the Officer's Club had no ice cubes at first. People could pull icicles off the eaves during part of the year and leave beer cans on the floor of the barracks, which was always about 40°F.

One night in the early days, there was no ice at the bar of the Officer's Club. The commodore of the convoy just in, the spring of 1952, was Jake Jacobson, who was in the Navy on the icebreaker *USS Atka*. He was disturbed about the shortage of ice, so he radioed the icebreaker to send a boat for him.

There were quite a few bits of sea ice and fragments of icebergs called bergy bits floating around. Jake got the *Atka* underway and pushed a good-sized bit on the beach, where a few officers with ice axes chopped it up and bore it in triumph to the club in jeeps.

The club had 10,000-year-old ice (10,000 years is the approximate amount of time it would take a flake of snow to fall into the ice cap and be compressed into ice that finally break off with the other snowflakes as glacial ice in the form of icebergs.) This ice is different from the ice made in a freezer, because there is air entrapped in it, and it's lighter than freezer ice. When liquor is poured on these chunks, they give off loud pops as the compressed air is released when the ice melts. The Officer's Club and NCO Club and the Airmen's Club all, and probably still do, brag that they had 10,000-year-old ice in their drinks.

The lack of refrigeration and freezing capacity was soon taken care of, and equipment was shipped to Thule. The story about the salesman selling refrigerators to the Inuit could be really true. The temperature is not always below freezing and cold enough, especially inside a building, to keep food preserved.

It might be expected that, with a thousand civilian construction people and large troops of Airborne, Air Force, Coast Guard, Army, and Navy people in a desolate land, where there was not a bush or a tree, there would be more personnel problems than at other sites. The old joke used to be, there was a woman behind every tree and, of course, no trees. But the civilian workers were carefully screened.

We, M&E, had a joint-venture architect engineer, and really checked our candidates for jobs. We did have a few soggy apples get in. But Mort Solomon, the colonel in charge and one of the real giants of Thule, ordered that no one would be hired by the joint venture who was not seen by Paton or Scheuren. After that, the odds tightened and our score approached 100%. The military had similar techniques, and the jail we built there was used for storage.

I was away from Thule most of the time with ski planes, dog teams, tractor swings, and weasels. Project Charlie Price, as I mentioned before, was devoted to locating AC&W stations at Thule area and also at other bases in Greenland, south of Thule. The first ski plane pilot was a reserve officer who flew for Eastern Airlines, a slim, blond boy named Kent Lawrence and a second lieutenant. He had mastered the technique of landing and taking off on ice cap snow.

The trick was not to stop immediately. The plane lost momentum, so you had to keep the plane going after touchdown so as to compact a little takeoff strip and also to cool the skis so they would not freeze onto the surface. The problem was later mitigated by the development of Teflon. This valuable material at this time did not apply to aircraft skis. We made the necessary flights to locate the stations, and Lt. Lawrence finished his reserve duty and left.

The pilots we got after that were not as skilled and resented advice from a non-airman, even if he was an expensive engineer. The next flight we made was with a green pilot. This time, we had some contractor people aboard to plan the construction at the site selected.

We directed the pilot to the spot. It is easy, when you were standing on the ice cap, to see an airplane overhead, but it's very hard for the crew of the airplane to pick out a dot or two on the vast expanse of snow. However, we directed the pilot to the spot. We had left some timber pile lines with red flags. The wind had torn the canvas flag to bits, but the timber was still okay.

The pilot, who had only an occasional use of one ski plane (C47) at the time, in spite of my suggestions, landed the plane, did not let it run very far, shut off the engines, and said, "Okay, get out."

We got out, set up some more markers, north and south, east and west, with empty oil drums and timber posts. Later, we found Christmas trees were the best. If the aircraft hit it, it would not tear up the ski. When we had finished our layout so the contractor could come back and set up camp and we could start the construction, I got a piece of rope, no bigger than a clothesline. I stood beside other people who'd gotten shovels, and I was going to saw the rope back and forth under the ski to break the bond with the ski and the snow.

The pilot said not to bother, he had a couple of cylinders of JATO (jet assisted takeoff). These were cylinders that blasted a jet stream like a jet engine and assisted the takeoff. The proper technique was to get the plane in motion and, after a suitable run, cut in the jets and lift off.

This pilot had not been properly trained. He got us all back on the airplane and, of course, could not move the airplane with the engines. He fired the JATO, which blew a hole in the snow, but the plane did not move. We radioed Thule, and soon a C119 flying box car, a powerful plane for airlifting cargoes and paratroopers around, arrived overhead and dropped the bottles of JATO.

Bill Moore, our chief surveyor, said he was going to write a paper on the physics: if the JATO bottle landed on the site, it penetrated the snow X feet; if it fell on the end, it penetrated Y feet. He could have collected considerable data.

It was quite a struggle to drag these JATO bottles in from where they dropped them. The first bottles turned out to be useless, because they could not fit the latches on the Air Force appliances.

The Army called it a C47, the Navy called it R4D, and Douglas called it DC3, while the British called it Dakota; each owner had developed different manuals and instructions. Quite an excessive expense, and it also ended up with us in a miserable waiting and camping out.

The nearest place they had JATO was Goose Bay, Labrador, so it was the next day before the correct bottles arrived. We civilians had all brought sleeping bags and so forth, but the Air Force had neglected to bring any rations except for a few chocolate bars. I dug a few snow blocks, placed them upwind of the hole I had dug, and crawled in for a nap. Finally, we got the correct JATO and took off back to Thule. Before this JATO was fired, though, the pilot was amenable to digging out the skis and getting the plane up to a reasonable speed before cutting loose the JATO.

Later, at another site (I was not present), the aircraft had mechanical trouble. There were no other aircraft available for ski planes, and our man Moore, the Corps of Engineers Captain Parcher, and the famous snow scientist, Dr. Henri Bader, were stranded for weeks. This trip had a bit of irony; of course, there was no booze, but they all wished they had some. There was some liquid in a barrel on the airplane that they thought was something to dewax their personal skis, but because of the altitude and temperature, they didn't recognize it was 151 proof rum. Anyway, it took a long, heavy campaign by Mr. Eddy to get all these people out of there.

We had a meeting in Wilmette, Illinois where the SIPRE (Snow, Ice, Permafrost Research Establishment) of the U.S. Army Corps of Engineers was located at that time. (It is now known as CRREL–Cold Regions and is located in Hanover, New Hampshire.) Bader used to say the laboratory should be located in Miami, because the laboratory simulated all the snow and ice and cold weather, and there was no reason to base in a place like Hanover.

I took my oldest son, Fritz to Wilmette with me. He had just started high school then and attended all the sessions in Wilmette then wrote part of my report for me. All the scientists there were very pleasant and called him "Scheuren, Jr." There was a Japanese scientist there, Dr. Nagaya. Nagaya was a real expert (although I hate to use that word).

Before WWII, the Japanese figured they would be conquering Russia, so they had studied and experimented in permafrost and snow and ice. There was no permafrost in Japan but plenty of snow in the mountains that causes much trouble to power lines. The Japanese had no use for their permafrost expertise. But Nagaya was a valuable consultant on the snow business in the ice cap. He described the densification of snow and later described the metamorphosis of snow into glacial ice.

Later, at Baffin Island, I was to warn Bob Peary and the colonel who was area engineer that to lay an oil pipe across a brook with no adequate support would cause a rupture of the pipe as the snow consolidated and dragged the pipe down. I keep a picture of that pipeline from my paper published by the Boston Society of Civil Engineers on "Arctic Construction."

Later, Fritz and I were to meet Dr. Nagaya at a camp on the Greenland ice cap at the Camp Century site where Walter Cronkite made a movie, *City Under the Ice*. We played bridge in a group in a Jamesway hut and listened to Nagaya's tale of a lake in Siberia that had frozen almost to the bottom. A hole was drilled through the ice cover, whereupon pressure that was locked in blew the fish out of the site into minus-50-degree air, and the fish were frozen before they could hit the ground.

In Greenland at that time, we had Henri Bader and Paul-Émile Victor. Bader was a Swiss scientist but a citizen of the U.S. Paul Victor was a French scientist and explorer who had done great work on the ice cap before WWII, in a way following the great German Wegener, who had died on the trail in Greenland. Paul is a great fellow who was in WWII as an officer in the Army, teaching ski troops in Colorado for the 10th Mountain Division.

Paul Victor, in the case of the first ice cap AC&W station, brought up the idea of submarine-type construction. That is, the weight of the station should be about the same as that of the displaced snow, so that station would sink in unison with the amount of snow accumulated.

We used eighteen-foot-diameter Armco multi-plate with half diameter dug into the snow and backfilled by compacting to the density of ten feet down or about forty percent. At these sites, Site 1 and 2 in 1953, there was not very much attention to Nagaya's theories as backed up the experiments. The exact load on a hollow cylinder placed in the ice cap and allowed to be covered with snow may (as Bob Phillipe said) not be closely computed, but it is greater than the weight of the snow directly over the cylinder, and the continually compacting snow tends to hang like a blanket or rug over a clothesline, as Nagaya said.

Bader brought up the action of a clay cut in a tunnel. The clay will tend to crush a timber placed across the arch. However, if the timber or steel beam is removed, the clay trimmed back and the beam reversed so that the sag had been raised like an arch; the tunnel can keep the same size, but a rigid frame would collapse. After the stations were completed, it took only a few years for distortion to take place. Putting in struts only aggravated the problem. Later, this was tested by monitoring tunnels in the snow.

However, during the construction of Camp Century and of the sites of Dye 2 and 3, it was difficult for the scientists and engineers to recognize the problem. I wrote to Al Crary, who was with the National Science Foundation, about the possibilities of putting a shear plane around the structures in the snow, so the structures would only have to bear the weight of the snow that was vertically above the frame. Neither he nor Phillipe would listen. Crary, of course, spent quite a bit of time on Island T2, floating around in the Arctic Ocean, and Phillipe was Chief Scientist for the Corps of Engineers, so they both should have known something about it.

When we built Dye 2 and 3, which were designed to be jacked up above the accumulation of annual snow, we had a foundation down thirty feet, where the snow density was about .6. We encased this foundation of timber, steel beams, and columns in a plywood housing and took care that the top of this was always kept above the surrounding snow, to preclude any trouble from blanket layer of the snow. But eventually, we had problems, because of the depth of the foundation and L/r ratio of some of the columns being increased so much, up 97 feet actually, but mainly because the ice cap is in constant motion, and while one station seemed to be moving toward the sea, the other station appeared to be in a sort of an eddy current that not only wracked and twisted the foundation members, but caused the communication dishes to veer from their assigned direction.

We expected these stations to have a seven-year life, but they have lasted over twenty-five years now, well over three times the designed life. The design concepts were developed at first in 1953, for Site 1 and 2, as a submarine setup; the later design, of course, was worked out for another style. In the early fifties in the Arctic, we greatly expanded our Thule concepts, and the Corps of Engineers improved the bases BW8 (Bluie West 8)–Sondrestrom, and Marsasuak BW1 and others.

The construction of 1953 radar sites was accomplished completely by air power. All heavy equipment and material were parachuted from C119 flying boxcars. Only personnel and sensitive electronic gear was brought in by the C47 ski planes. The airdrop was successful, but there were a few streamers when the chutes failed to open and the material fell free.

The weather wasn't always clear, and when we had to drop through an overcast, we used to rig the drop zone with people. Radio was rough. The planes would approach, guided only by the answers of the group on the ice cap, who were in turn getting information from the noise of the engines. The people on the ground would point to the area where it appeared the airplane would be approaching, and the director would use all these visual and audible aids and then signal over his radio, "*Execute,*" and the stuff would come raining down.

One large 16-inch-wide flange beam slipped from its chute and plunged into the emergency hut, where clothing and food were stored for use, if the main camp burned down. It was always a hairy procedure; it was not used unless we really needed the stuff and no better weather was predicted for the near future.

I remember one day when Andy Paton was visiting, several chutes "streamered," and one chute came loose and drifted over the mess hall stack, so we had a little fire. Andy coolly asked if this was a regular noontime procedure. There was no booze at the sites, not even beer, which was allowed in Thule. But the morale in the main remained very good. We had a snow melter and hot water, so showers were available.

In another case, the specifications called for insulation of the diesel generator to be formed from diatomaceous earth. When the material, which was of course preformed, arrived, it was found that the insulation was not as specified. It was magnesia.

A weird telex came in from the site describing the difficulties of wrapping the pipe with this earth and that there would be no room to have the earth and preformed magnesia insulation. Use of the diatomaceous earth material was a refinement that the designers wished because of the high heat generated, but we could not wait for the material, so made do with the magnesia. The site superintendent, the site architect-engineer, and the Corps of Engineer residents at the site had no knowledge of the material.

A few hundred miles south of Thule is Sondrestrom Fjord, an airbase called BW8 Sondrestrom AFB, which was built during WWII, 1940-1941. The runway was 5,000 feet long, built on a terrace along the Watson River, where it enters the fjord. This 5,000-foot-long terrace was practically a natural airstrip, except for the clay material that needed considerable treatment to hold heavy aircraft.

In 1928, for instance, Parker Cramer had taken off from here to try to fly to Europe. He ditched on the ice cap and made his way back to where Professor Hobbs of the University of Michigan had a geology camp. His co-pilot was the late "Fish" Hassel, who was later in on the development of Thule. A long mountain paralleled the runway at BW8 named Mt. Hassel. The great flyer and explorer, Bernt Balchen, who was the first pilot to fly over both poles and who flew Byrd over the South Pole, commanded the base here during WWII.

By 1952, the base was still rough and ready. The runway was short, and a mountain called Sugar Loaf interfered with the use of one end, so that field could only be used by flying the fjord.

Our project was to build a base equivalent to Thule. We developed a water system by use of Lake Ferguson, which was a fine source of good water. We put in a sewage system that discharged by wooden pipes into the fjord. We extended the runway and built additional barracks and a system of POL (petroleum, oil, and lubricants). We also surveyed and made a preliminary design of Prices Peak, a proposed AC&W site named for the great Charlie Price. It was never built. This port here was a pier at the end of a dredged channel through the outwash plain of the Watson River, and the POL was brought in by undersea pipelines. We updated the POL system, added undersea lines, and built tank farms at the port and pipelines to the tank farms at the airbase.

We had a 1952 look at the problems of doing government work. There was a rocky point about a mile and a half along the fjord from the existing port facilities. I had been on the *North Wind* icebreaker for several days with their Greenland Cruiser and had sounded the area at this point, which was called First Point. This appeared to be an excellent place to place a dry cargo dock and also a POL dock.

We did a preliminary design on such a facility, which would have allowed cargo ships and tankers to come alongside and discharge cargo to storage and to truck, then hauled along a mile and a half of road to the existing port area. The tankers could come alongside and pump directly into the pipelines to the tank farm of the airbase. This facility would thus eliminate the need for undersea lines and also the need for the lightening operations required by hauling barges alongside ships anchored out in the fjord and then towing the barges up the dredged channel.

This required frequent dredging and transferring the cargo a second time to trucks. At that time, because of the delays to shipping, it was costing Uncle Sam about $500,000 per year in demurrage charges alone. The new dock and road would have cost about $3,000,000. The Transportation Corps complained that the existing system, expensive as it was, was valuable in training the troops.

Congress is happy to waste money on operation or maintenance but does not wish to finance what they call capital expenditures, even if these save money. So, this improvement died. Later, when it was proposed to extend the existing 5,000-foot runway to 9,000 feet over the clay terrace, we found that, if we had set the runway out along the outwash plain, we could get 10,000 feet on top of the 160-to-170-foot depth of well-graded gravel and also get a runway that would work from either approach. Besides, all the construction could take place without any disturbance of the regular activities of the existing runway.

This time, we designed both alternates and advertised it that way. The low-bid construction contractor offered almost $3M less to get the 10,000-foot runway that would operate both ends than for a 9,000-foot runway that would still be only a one-way operation and subject to settlement and the construction of which would disturb daily operations. However, the lesser value and greater cost was accepted.

The 9,000-foot runway has required excessive maintenance. All the material hauled to make the required fill was obtained from the area where we proposed to build the 10,000-foot two-way runway. One argument used was that the extension of the existing runway would have justified the building of the 5,000-foot strip in 1940. Of course, these people avoided the truth that, at the time, 5,000 feet was plenty. We had a similar struggle many years later when I wanted to build a new strip at Kulusuk on the east coast, rather than use the Bluie East 2 existing WWII base at Ikateq.

While we were working at BW8, there appeared to be a need for a relatively temporary 10,000-gallon barrel tank for aviation gas at the base. There was such a tank at Frobisher Bay on Baffin Island, and it was proposed to move this tank to BW8. A Corps of

Engineers civilian, Joe Russo, was dispatched to Frobisher Bay to check on the shipping of this tank. When he arrived, the ship had already left with the dismantled tank.

He sent a telex: *Mission accomplished, Russo.*

It had not really been accomplished; the ship left without the neoprene gaskets for the tank erection. The tank, in spite of our objections, was erected without efficient washers, etc. I told the area engineer, Joe Bacci, that I wanted nothing to do with the erection of this tank. I would design and monitor the building of an adequate dike around it, to hold the oil that would leak out. Of course, the tank leaked, and some idiot told the major he should get a diver from the States to dive down into the AV gas and caulk up all the seams. He was about to send such a message, but at my almost insulting urging did not do so.

I climbed up the ice cap on the road reconnaissance to find the best way to build a road up through the crevassed zone to the stable ice cap. Al and I climbed around. Later, a Transportation Corps pilot saw our prints and said that we had crossed a snow bridge. He could see the openings of the crevasse at each end.

We had ropes, ice axes, and so forth, and we had tested the bridge and walked across it. Al, who weighed about thirty pounds less than I, always claimed he was going to cut the rope if I broke through. Later, he resigned from the Air Force and wrote articles about the foolishness of building Thule.

Anyhow, our snow bridge got Bacci excited, and one day, he and a batty Transportation Corps major got some skis and started to ski to an island off Thule. This island was farther than it appeared, and the officers got lost. Before they fell into the open sea, they met two Inuit with a dog team who were heading for home. The dogs knew their way, and although the Inuit couldn't get through the whiteout of the snow fog any more than Bacci and his friend could, the dogs got them all back to the base. Bacci spent a few days in the hospital for frostbite and exhaustion.

We had to handle the rehabilitation of Narsarssuak and Siniurak BW3 on the southwest coast of Greenland: very south, no permafrost. The weather here was always much worse than anywhere else up there, so bad that even the sea gulls could not fly half the time. It was the base BW1 from which they still flew to Iceland and Goosebay, delivering airplanes to the U.S and Europe.

During our first days at BW8, there was a small Danish presence there. At BW1, there was a Danish hotel and a large group of Newfies, Newfoundlanders, who worked at the base.

Our people, as well as the contractor's people, had pretty primitive quarters at the time at BW8. But at BW1, the conditions were worse, and we had some trouble with some of our people, the Corps, and the contractor. Our manager here turned out to be one of the bad apples. Instead of encouraging his men, he really led them in complaining and bitching. We got rid of him and brought old Lou Litchfield out as manager. Of course, Lou got the morale part straightened out.

It is always hard to be an architect engineer with an agency like the Army and Navy and so forth, who have many engineers working for them. The field people, the area engineers have one idea; to get the job done, the AE may be suspected by the area engineer for holding up the progress, and the contractor gets more license than the AE may wish to give.

Little used to beef every year when he came to Boston to renew his contract for the next season. I used to take him to the Engineers Club one year, and he would take me to Jacob Wirths the next. There were several things that either Paton or I had overruled him on. Lou would say, "You said let them go, so I did, but it wasn't right."

Our chief of civil design in Boston, Nat Clapp, and Lou Litchfield were at the airport to welcome me to Thule in early 1952. They each wished to, and did so at different times, show me two culverts built under a taxiway. One was only about a foot and a half deep, and the other was about seven feet deep.

During the 1951 season, there had been a bitter argument between them concerning the depth. Mr. Eddy remarked at the time, "The two most stubborn men in the world," so both pipes had been built.

Each had a ½-inch pipe within it, so that during the spring thaw, steam arose from a portable boiler, called a prospector's boiler, which was used in Alaska by prospectors to wash away frozen soil in the search for gold. So, both Nat and Lou wished for my criticism, good or bad.

Actually, I had to agree with them both: the upper caught the earliest flow, and the lower one caught the later flow. They both were useful, and Nat and Lou became two of my dearest friends. Both of these engineers were real giants in building the Thule Air Base and the work at P Mountain, AC&W stations BW8, BW1 and BW3.

When we returned from the Antarctic, we found ourselves in the midst of the BMEWS (Ballistic Missile Early Warning System). There were three sites for these. One at Thule, one in Alaska, and one in England. M&E was architect-engineer on the first two places, not the one in England.

We had set up a special office in Boston. These offices were like a zoo: crowded, noisy, and with meetings attended by the Air Force, the Corps of Engineers, M&E, RCA, Western Electric, the Telephone Company, Sylvania, and so forth. There were about sixty participants, so it was impossible to get anything done.

I went to one meeting and walked out. All of the sixty seemed to have one desire, which was to get their names in the minutes of the meeting and to write memos with all sorts of security classifications on them. We were particularly interested in the proposed layout and the space required. Somehow, we all stumbled through, and we built the place with no huge structures left over, nor structures in which floor space was crowded–lots of luck.

I couldn't stand the crowded conditions in the new office and moved back to a little space at M&E's main office building. I remember a paper I wrote on the first ice cap stations we had built, which had been added to by a civil servant with the Corps of

Engineers and sent back to me under heavy security. That was back years ago. Because I had no place to keep it, at that time, I just sent it back to the idiot who'd shipped it to me.

Sure enough, some character from RCA here in 1958 sent us a lot of garbage, heavily classified, criticizing our design of using timber poles to carry power from the site to the post power ship for temporary use until the new power plant was built. This guy claimed that the icing conditions would be so bad, the cables would break and the poles would break, so we should build a tunnel from the ship to the plant and so forth.

From 1952 to 1958, I had measurements of icing and hoarfrost in the Thule area and a good estimate of the icing on the transmission lines and poles of about 3/8" of hoarfrost and a specific gravity of about 2.7. In this report, there were photos of telephone poles with cross trees buckling under the weight of ice. They couldn't have been taken at Thule, because no such poles with cross trees were ever used there.

They probably were taken in Baffin Island, which is another story entirely. There, we found twelve inches of radial ice on some of these pole-type metal antennae. When we broke this ice off, we could put the segments to fit a 24"-diameter barrel and use it for drinking water in our housing there.

This pseudoscientist had a great deal of mathematics and so forth, but when you start with the premise that 2 x 2 = 5, the rest of your mathematics may be impeccable, but you will never be correct. I sent the whole report back to RCA and forgot it. Later, I found this writer had figured the moisture in the air by consulting a so-called meteorologist who could tell the amount of moisture in the air by visual means. Anyway, no tunnels were built.

During the construction, a power ship was set in a basin protected by riprap and so forth from the ice action in the bay. Some ice formed around the ship in the enclosure, but there was no harm. There were four steam lines leading from the ship's boiler to feed the need for heating. These four lines were flexible to move with the ice and rise and fall of the tides. They tied into a 20"-diameter header line that was anchored to a heavy concrete block. It had a 2″ valve on its lower end to drain it. There was also a condensate return line, 4″ in diameter, that carried the condensate back to the ship's boilers.

The steam lines and the condensate return lines were insulated and lay on the ground. When the steam had been condensed in the heating process, the condensate was returned by steam traps that kicked the condensate out of the steam lines and into the return line. This required maintenance. Bypass valves were provided so the steam traps could be checked and, if faulty, could be replaced with rebuilt ones and then taken to be repaired.

Burns & Roe engineers had a contract to operate the power ship and the Base civil engineer from the USAF was supposed to maintain lines. One night, the header broke loose from the welded fittings that tied it to the anchor block. The header split, and the steam and water came out, and the ship was shut down.

There were several telexes sent to Washington, all blaming M&E for improper design and so forth. When I arrived, I found our people were checking the block for settlement, growling about the quality of the weld and so forth. It took a little time for it to sink into my thick head.

John Napier, may the good Lord rest him, was a great bowler, and besides bowling on the M&E team, he also bowled on the Burns & Roe team. He came to me and said, "I don't know if this is of any value to you, but one of the bowling teams said that, before the header split, it was bouncing up and down."

Well, he had solved my problem: when the ship shut down for one reason or another, the drill was to open the drain at the end and drain any water still in the pipes that had accumulated there during the cooling of the steam. This had been done. That is, the valve had been opened but no water came out.

We found that, by taking a piece of rebar and ramming it up the open valve, the valve was plugged with scale and would not drain. Then, the steam traps were not all working, so the condensate, instead of being kicked out in the return line, just stayed in the pipes and, when the ship was shut down, ran back into the header. The header was full of water, so when the superheated steam hit it, there was no place for the water to go. This caused a water hammer, which made the header jump around and break.

Napier's remark had, of course, indicated that the water hammer was the cause, so I called a meeting at the ship. The base commander and his engineers were there, and I sketched what had happened on a blackboard.

The base commander asked what I could recommend. I said to repair the damage to the header; make sure that drain is operative, put in a larger valve, say 4", and repair the faulty steam traps. Constant maintenance, too—have two men continuously checking the traps and replacing and repairing them. There were no more problems after that, and the power ship operated until the permanent power was installed.

A small sidelight on the BMEWS operation: when the power was turned on, a target appeared, and the red button was almost pushed. However, it turned out to be a rising moon. All were looking at the screen, and no one looked out the window. The computers were adjusted, and then this was okay.

The second thing that happened was, the General Electric representative, who had been complaining all the time about the unnecessary expense of sealing each weld and each joint in the complex from radio frequency, climbed up the side of the antennae and down the other side and across the top without being microwaved for lunch.

Ken Lippit, a friend of mine from MIT, told me one day that the amount of RF radiation, except right in front of the feed horns, would not be dangerous. The passion for this RF protection was so great that, when I returned from Antarctica, I found we were about to recommend that the Corps approve a contract for well over a half million dollars to be paid to an electronics firm for a system to check on this seam welding. The instruments that could measure it during construction had not yet been invented.

I vigorously opposed the idea, even though our chief architect said, "We've been studying this for months, and you've only been here a few days, and you say it's no good."

I took him with me up to the Air Force Suppression Branch at Rome, New York—all airplane cockpits and flight decks are supposed to be closed rooms—and down to Fort Monmouth in New Jersey, home of the United States Army Suppression Branch. This suppression means protection of things like protection of radio and so forth, through proper shielding like spark suppressors in an auto engine.

A German immigrant, Bruno Ripke, a delightful person, went over the matter with me, and we found there would be no use checking until the enclosure was complete. During the construction, we assigned a fine young architect to monitor the welding, and when all was closed in, Mr. Ripke sent a captain and three enlisted men from the Signal Corps with the ordinary monitors. The total cost was less than a tenth of what the useless proposal would have cost.

Ripke was a master of what appeared to be double talk, but it was solid philosophy. I had him come to Thule and lecture our people, the Army people, the Air Force people, and construction people. He asked me afterward if I thought he had confused them enough.

The final job was successful, and we had provided adequate space for the equipment and left no areas not filled nor any that were too congested. I thought this was success, particularly when the equipment was being designed while we were getting the station built.

Concurrently with the BMEWS project, we were engaged in the extension of the DEW Line, which is an electronics fence across Canada and hooks on in Alaska to the White Alice job. We were also involved in the DEW Drop; this was to be a connection of the DEW Line at the eastern terminus and the base at Thule.

Several routes were proposed and required study. One route was an intermediate station near Clyde Inlet on Baffin Island, near the Coast Guard station in Cape Christian, which we had built years before for a Loran station, a navigation aid. Another purpose was to use a hill near Pond Inlet, which was on adjoining Bylot Island. The third method was to shoot all the way from Dyer to Thule, which of course was farther than the 440-mile limit for Tropospheric Scattered Transmission. And lastly, we were to investigate the use of an undersea cable. Around the same time, we were assigned reconnaissance missions to establish stations on the proposed DEW Line Extension from Cape Dyer, the eastern terminal, to the west coast of Greenland; possibly two stations on the ice cap; a station on the east coast; and possibly one near the southern tip of Greenland.

The eastern coast of Baffin Island is only free from ice for shipping, even with icebreaker help, for six to eight weeks in the summer. Access to Cape Dyer and Cape Christian was, at that time, only available by ski plane operation at Christian and by a very short runway along the side of a cliff at Cape Dyer. The ski plane operation meant you had to land on the ocean, on frozen sea. A larger strip was being built at Cape Dyer, where four-engine planes could use Frobisher Bay, which is a joint US/Canadian base that had been built in WWII and was now being refurbished.

We therefore sent a party in a ski plane from Thule to Bylot Island. We sent our resident manager at Thule, Sandquist, with this group. He could bring no other survey people with him because of the Air Force, Transportation Corps, electronic people, Canadian Navy, and so forth. Only Sandquist and Commander Crowell R.C.H. made the ascent to the icy ridges. The access to the peak was so difficult, the site was dropped from consideration.

I was at BW8 then. I had already flown in a ski plane to Cape Dyer. The DEW Line contractors were extending the base there. I did get enough information so I could instruct a survey party for a later survey.

The Canadians there did not know much about permafrost and its problems. While we were there, their chief engineer asked me to check the excavation for the tower construction they had started. I had on board a prospector's boiler with which to thaw ground for samples and so forth. However, we did not carry it with us to this site.

When we arrived at the excavation site, I found the site partially excavated but with a large group of men with picks and shovels trying to dig up some unexploded dynamite. In cold weather, the battery or generator that is pushed down by hand to explode the charges does not usually explode the fifty caps for which it is rated.

In this case, they had drilled fifty holes, loaded them with dynamite, and tried to fire. Only a few went off. I told them that their method of digging was dangerous and offered the use of our prospector's boiler, suggesting they get a compressor, which they had on hand. By melting the frozen ground, they could carefully blow out the stemming to get it to powder, so it could be refired.

I refused to walk across this self-rigged mine field. The Canadians said, "You Americans make too much of safety. You have to take chances. I've been hurt several times myself."

Anyway, I picked up a few pieces of frozen earth and took them back to the tar paper shack we were living in and put them in a #10 can and melted them on top of the stove. They proved to be about three-quarters ice. Obviously, they must excavate deeper and place non-frost-susceptible material of sufficient thickness for a foundation.

Then, I returned to BW8 on a SA16 Grumman Albatross Amphibian. It was quite a flight. We had lots of icing, so we had to keep dropping down to escape it. The fuselage was getting plenty of whacks as the ice broke off the propellers.

Western Electric, or rather a branch of it, Federal Electric, was operating the DEW Line. Federal Electric had a great manager in New Jersey, an admiral named Cruzen, who had been with Byrd in Antarctica and had lost his son, a Navy pilot down there. Western Electric had given more topography to the Corps, and M&E had taken it and used it without attribution on some drawings.

When I sent a survey party to do the layout for the contractor, our chief surveyor there, Herb Yamane (MIT, then USAF reserves, then M&E) and his party found out that the topography was incorrect. One of the stations would have to be hanging on the edge of a 2,000-foot cliff. We had to redo this survey and got blamed for it, because our designers had not sense enough to mark the plans "topography by Western Electric" or "topography by others."

The reconnaissance for the other alternate on the DEW Drop to Thule, Cape Christian, was slow getting off because of the unavailability of ski-equipped aircraft. When we finally took off, it was getting late into spring, and the ice was melting on the east coast of Baffin Island. The usual problem occurred: the lack of space on the airplane for me to bring along any help, neither surveyors or geologists. The airplane was filled with civilians from the Corps of Engineers, Air Force, and from the electronics companies.

One of these was a fellow named Mains, who knew something about the needs of Troro scatter transmissions. The others were along for the ride and hope of hunting polar bear. Besides the air crew, there was an Air Force major, Will Blair, who was a fighter pilot and a bit overweight, who had been assigned as an Arctic expert because of his service in Alaska. The Corps of Engineers was represented by a civilian, Arthur Alexander, who was later to be with us on the reconnaissance cruise on the *Westwind* and later the survey crews of the *Arctic Sealer*. A couple of other civilians were along from various government agencies, as well as the Transportation Corps Captain, who was along to check the beach.

The crew of the C47 consisted of Major Steve Oliphant, Capt. Killer Kane, and Capt. Joe McGuckin, who was the navigator. I could not bring along an instrument man, but I could use McGuckin to run the theodolite (a surveying instrument with a rotating telescope for measuring horizontal and vertical angles). He quickly checked out and was an excellent operator.

We landed the airplane on the sea ice in the bay near Cape Christian. We were met by Chief Petty Officer Hatch from the Coast Guard and Corporal Doug Moodie of the Royal Canadian Mounted Police, the Inuit Simon and Simonee, and two dog teams. We traveled to the Coast Guard Loray Base at Cape Christian.

The rest of our party were a bit upset because the Mountie invited me to share his house. They did not think it proper of me, my rank, to move in with a noncommissioned officer. I told them my friend, Corporal Moodie, was Corporal in the Royal Mounted Police and that ranked at least as a Major General in the US Air Force. Moodie had a very comfortable two-bedroom house. The rest of the party stayed at the Coast Guard quarters.

Both possible sites were a few miles away and, when we arrived, were clouded with fog. I set up the theodolite light over a marker our surveyors had placed there two years before and said, when I could see the top of the mountain, we would travel with the dog teams, no matter what time it was.

There were two Inuit families there, and Simon and Simonee and their families lived in these two tents. Their dogs were chained up outside. Because the airplane was already gradually sinking into the melting ice, we did not have much time.

After a couple of days, I still could not see the peak. Moodie and Chief Hatch whispered to me that the rest of the party were getting a little itchy to go hunting and so forth. I reiterated my decision to watch the peak and, when it really started to clear, to get the two dog teams and head for the hills.

Finally, it cleared about 10:00 p.m. one night. Of course, it was twenty-four hours of daylight there. Then, the excuses started. Some did not want to leave until morning; some said that, after all, they didn't need to go so long as I went. Finally, all who went besides me was Capt. McGuckin, Mains, Major Blair, and the two dog teams with a few cans of C rations we never used.

On the way we stopped in to visit the factory of the Hudson Bay Company, which had a trading post at Cape Clyde inlet. It was a very comfortable place, and Main thought we should stay there for the night. He said the dogs might need to be fed. For a less than twenty-four-hour run, neither the dogs nor we needed either rest or food, so we pushed on.

We only rode the sleds across the sea ice. Once we had to climb, we took the dog sleds along to carry the theodolite and such emergency rations as we had. I laid out the routes for possible roads. Mains and McGuckin took one, and Blair and I took the other; each with an Inuit and a dog team.

Snow had started to melt and then froze, so the crust broke through sometimes. It is tough going when the crust breaks at each step, and it is better of course if the crust does not break at all, but when the crust breaks part of the time, it always seems worse going.

Blair kept falling behind; his Arctic experience was only in the cockpit of an airplane. He moaned that it was embarrassing to have an old man, meaning me, breaking trail for him. I told him to go ahead, but then I would be soon breaking trail again. He was around thirty-six, and I was about forty-nine.

We stopped for me to check a possible switchback in the road. There was a large, flat boulder, snow free, and Blair lay down and was happy for a few minutes. Said it was the softest rock he'd ever lain on. We met the others at the top and McGuckin set up the theodolite to check the terrain toward Thule and also toward Cape Dyer for any interferences with the line of sight. Each komatick (sled) had a polar bear skin, and when Simon pulled one of them off and wrapped himself up in it, McGuckin said he then began to feel cold. We returned to base.

When we got down to the sea ice, we climbed onto the sleds, and the dogs needed no urging to gallop home. We had been gone about eighteen hours and traveling all the time, except for the observations at the top.

We flew out not an hour too soon. Major Oliphant had checked the aircraft, and there was six inches of water around the wheels. He had to consider leaving us behind to save his airplane. We could have made the trip much earlier when the ice was harder, but as always, the people at headquarters, civilian or military, are not really understanding of the plight of people in the field. Anyway, the final decision was to lay a submarine cable from Cape Dyer to Thule, so the trips to Cape Christian and Bylot Island were unnecessary.

The next sets of reconnaissance were for the Eastern Extension of the DEW Line. Map reconnaissance had identified a peak on the edge of the ice cap at the head of a fjord, west of BW8. This peak, called Qakotokaq or "biggest hill in the area," was about 5,000 feet high and practically due east of Cape Dyer. The fjord, which ran from Ivislak on the west coast of Greenland up to the base of this hill was called Ekalugarsuit Ekaluk, Inuit for salmon trout—what we call char. The whole name means the "place where there are many trout."

I went there for the first time with a young Danish engineer the year before. We landed near the base of the hill and climbed up, about eleven or twelve miles on a ten-percent grade to the top. I had had some hand engineering instruments like an Abney level (a surveying instrument that measures angles, grades, and heights), and in those days, it was practiced in walking ten percent until the switchback was needed, and then starting ten percent the other way.

We located a few places for gravel quarries and so forth and some small ponds along the way for a possible water supply. I remember finding two sets of caribou horns on the top of the hill. The last 100 feet to get there were practically hand over hand over large broken rocks. The top flattened off toward a permanent ice field at the edge of the ice cap. It seemed hard to believe that a caribou would have climbed up all that rock and ice when there was certainly no vegetation or food. Later, Mark Klinefelter, who was our resident chief engineer at BW8, spent more time there and laid an approximate road location (which we later surveyed and built).

Then, we were given a chance to spend two months on the Coast Guard icebreaker, *Westwind*. On this trip, we traveled from St. John's, Newfoundland to Greenland and all along the Greenland coast, east coast up to above Scoresby Sound, and looked at about thirty-five possibilities for the proposed stations to tie into the DEW Line from the proposed ice cap stations to the new bases in Iceland. This voyage started the same way as all the others:

I had to make the study, but I wasn't allowed to bring any survey help, because the extra passengers, call them tourists, both civilian and military, crowded the ship.

I did manage to get one of my people with me, Dick Call, about twenty years old and son of our chief engineer at Thule. He was studying for his PhD at the American University in St. Louis, and he was an excellent instrument man; as a budding geologist, he was a great help, of course, on the reports. Dick went back to school and got his doctorate and now operates his own geological engineering firm, specializing in mining engineering in Tucson.

The *Westwind* was commanded by Capt. Peter Smenton, who later became a professor with the Coast Guard Academy. Pete was a very erudite gentleman and expert seaman and had long experience in the Arctic. He had been with "Ice Berg" Smith in the early 1940s. His executive officer was Commander Ken Ayres. The last time I saw Ken was several years later in New York, when he passed through on his way to South Africa to enter Antarctica via the Weddell Sea.

The engineering officer, Lt. Rodgers, was also erudite, and besides his expertise in the operation of his engines, he was very skilled in semantics. At the time, he and I were talking about Nancy Mitford's book, *Noblesse Oblige*—U and non-U speech and so forth. These three and I had, for two months, an ongoing bridge game. Each night, after the evening meal, we retired to the captain's quarters from 1800 to 2000 to play bridge.

At 2000, we went down to the ward room to see a movie. These were not always the greatest, but it was a matter of form to attend them. Following the movie, we went back to the captain's cabin and played about two hours more, and at 2400, the Captain's Steward served coffee and small sandwiches, and we went off to our respective rooms to sleep.

These bridge games went off no matter what the weather. The only consistent thing was overbidding. The main theme of "down one" was frequent.

The Danish liaison officer on board was Commander Schou. He looked like Prince Philip of England; and he possibly had some relationship to him. He used to watch us play sometimes. Once, he filled in but never again. He couldn't stand the down one bit. When the ship was not in the ice, it used to roll like a bottle on heavy seas. The table was bolted to the floor, and we had lanyards to tie our chairs to the table. The dummy had to usually hold both hands on the cards to keep them from sliding off. The dealer took off the shuffled cards and put the deck in his breast pocket.

Captain Lally was a brave officer, but he was afflicted with sea sickness. When you are seasick on a ship like that, you get no sympathy, even if you're sick all the time, and the waters in Denmark Strait between Greenland and Iceland (this was in the fall) are always rough. Besides the ship's officers, the captain ate by himself. We had several Air Force, Army people, officers, and civilians who ate in the Ward Room. Most of these people were seasick and couldn't eat much. Commander Ayres, who was president of the mess, used to crow loudly that the mess was making a lot of money because all these people were paying mess bills and could not eat much at all.

I've always been fortunate not to ever meet up with sea, or air, sickness. In spite of the weather and some seasick people, we managed to visit over thirty possible sites and finally designated three of them for future survey. Really, the only ones who had any work to do besides running the ship were Capt. Lally of the Air Force and the electronics communication man, Dick Call, and Capt. Tatch, who was a young engineer officer and myself. The others were merely along for the ride.

We visited Greenlander towns or settlements all along the way. We had two helicopters on board. One belonged to the ship. It was a two-place Bell, a bubble of plexiglass. Capt. Smenton called it a magic carpet. And so it was when I rode with Emmett, the pilot and asked him to take me on various routes to the peaks we were checking.

I could tell in a few minutes what lay in wait beyond the various ridges. To do this on foot would be necessary at some point, to make the proper survey, but to get the information we needed now in a few minutes in a helicopter saved many days of climbing and enabled us to discard many of the sites that had looked plausible on our map study.

Capt. Smenton had developed a quick way to get the helicopter back to the ship, and I accepted this. It seemed a bit hairy at first, but my role in flying is always to trust the pilot whenever we were in the air. Emmett would cut the engine and rotate without power almost down to the deck, then he would cut the engine back in. This was the quickest way down.

For longer trips, we had an H19 Air Force helicopter that could hold eight or nine people, except for the pilot and co-pilot. The visibility in this aircraft is limited, and although the H19 saved us lots of walking, it was not as good as the Bell for quick reconnaissance.

One of the sites proposed for investigation was Stur Island (stur means big in Danish). We had pretty good aerial photos of the island, and I had laid out a proposed route to the peak. This route climbed from the beach to a switchback and turned left, and then it looked feasible to cross a sort of terrace that appeared to be competent rock. At the end of the terrace switchback, turn right and continue.

We had several days of bad weather before this. Once we arrived at Stur, we were getting behind schedule. On this trip, as well as the next year's cruise on the *Arctic Sealer,* we found ourselves getting so late a start in the fall, we had to cut our time to suit the approach of winter.

There were three or four other sites that we wished to look at in the area, so one day, it was decided that Capt. Tatch (he was a young civil engineer officer) would go with my sketches of Stur Island and check this site. I would go on the H19 and take a quick look at the other sites. This took me all day, and none of the places were feasible for our use.

Stur, it turned out, was not, either. Tatch climbed up to the proposed site of the proposed switch to the right of the terrace, but when crossing the terrace, he had found out that most of it was covered by a sheet of ice formed by water seeping out of the cracks in the rock just above the terrace. He made the ascent and brought back his report.

I can always remember his comment about crossing the icy terrain on the edge of the cliff, when he described this part of the ascent. He said, "Even you (meaning me) would have hesitated crossing this area." And probably, I wouldn't have crossed it at all and would have felt that this terrace could be written off and not feasible for a permanent facility. Tatch completed his mission like a good engineer/soldier, and we discarded this site because of the difficult access.

When we were at Ft. Pepperell, St. John's, I looked in on my old friends, Jack Crowell and Bernt Balchen. Bernt had just returned from a sealing ship cruise along the east coast of Greenland. He gave me some photographs of some areas we had intended to visit, especially the Scoresby Sound area. He also gave me some locations of some caches of aviation gas he'd left behind for future helicopter use.

It was sad that Bernt could not come with us on this trip, but he was about to be retired (Col. US Army Air Corp. USAF, now; Air Corps when he was in the war) and had lots of paperwork. Bernt requires more than a mention in this chronology, and I will write about him later.

Scorseby Sound had a radio station operated by the Danes and Greenlanders. We visited them, and after our investigation of the Swainson area, headed south. The Angmagssalik area had been developed in a small way by the US in WWII. A weather station and radio station had been set up at Cape Dan on Kulusuk Island, a little north of Angmagssalik. Angmagssalik is a small Greenland town in a short fjord or bay in from the sea, near the edge of the ice cap.

There was a radio station there, also. One of the Danish operators, Thane, was a friend of mine from Thule. He was an old hand in Greenland and had his wife and children with him. His wife loved to travel on the dog sleds.

I remember her telling me in Thule once, with stars in her eyes, how wonderful it was to go out on a trail for several days. "Fine, fine," she would say about sledging and sleeping out on the sea ice. We stopped in there several times to send radio messages to headquarters.

I had another friend there, Neil Underberg, who operated a little ship called the Miki. He used to move Greenlanders about and carry cargo. He had some small barges he used to tow with loads of sand and gravel and cement and so forth. He was a good artist, and I have a painting of his showing his Miki in the foreground and a large iceberg with a hole in it. This iceberg was grounded near there. A picture of looking through it at an icebreaker is in the 1958 Practical Navigator.

Once we were in port there, our C119 took a few Greenlanders for some short hops. Neil and his wife, Kitty took a ride and left the two children (ages two and four) in my care. The helicopter disappeared toward the ice cap, and I stood there with a two-year-old girl in my arms and holding the hand of a four-year-old boy.

An old Inuit smiled and told me, "If they do not return, you'll have to take care of those children."

We also used to take the Greenlanders, both Danes and Inuit, on board the ship for movies, soft drinks, and cookies. Of course, there are old tales of Inuit offering their wives and so forth, but I do not believe there is any more wife swapping among the Inuit than there is among the small towns in the U.S.

However, the Inuit does not hunt like the Kabloona (white man). The white man goes out, all male, and travels around; no one cares for the cooking, clothes, and so forth. In the Inuit, the wife is a solid member of the team. The man hunts for the winter meat, and the woman travels with him, staying in the camp, snow house, or tent; she keeps the lamp trim, cooks the food, and repairs the clothes. The clothes are very important; in bad conditions, the clothes can mean suffering, and in good conditions, happiness. When the man returns from the hunt, the igloo is warm, there is hot food, and the clothes get repaired.

Anyway, the first time we had some Inuit on board, some of our people thought all Inuit women were fair game. They soon found out that, while the unmarried girls might be more or less available, if they were to approach a married woman, the other women would point toward the woman's husband and say, "No, no."

That wife sharing seems to be uppermost in many people's minds, but when one considers the inability of the Inuit to lie or steal, as well as to be unable to understand why men would gather in groups to make war, I believe we should devoutly wish we could all be like the Inuit.

Besides the radio and weather station at Kulusuk Island, the U.S. had built an air strip at Ikateq. This was built on the side of a fjord north of Ikateq and west of Kulusuk. The strip is hard to find from the air, and ships can only reach it a few months of the year.

I remember showing some slides of the area to a group of officers and civilian Corps of Engineers of the Northeast District and being accused by one of the colonels of taking pictures of "scenery," until I got them to see this little gravel terrace along the fjord, with the northern end shadowed by a 5,000-foot peak or Sugarloaf. There was a better place to put the strip on Kulusuk, where we did build Kulusuk International Airport.

I never asked Bernt Balchen about this, but I believe the WWII people recognized Kulusuk as a good site but chose Ikateq for camouflage purposes, because there were German weather people, submarines, and so forth up and down the coast. I had read about the Greenland Sledge Patrol, some Danes and Greenlanders who traveled along the coast to check up on a possible German invasion, and I'd met Hansen the Dane and another member of this group who was in Scoresby Sound when we were there.

The Inuit on the patrol were really puzzled when a group of Germans killed one of the Danes; as I have said, they cannot understand war. If Greenlanders have now become more warlike or formed a state, I cannot say. What did Hemingway say: "War is the health of a state"?

We looked at Ikateq to get a picture of what could be done if we used this site as a supply point for sea and air lifts. It seems clumsy, but the field existed, and I already knew the political problems of a new capital expenditure instead of changing something else that was already built. Then, the helicopter (H19) required a part. We had been at Narsarssuak for a visit. We were not really welcomed by the Air Force people there; we'd had a group there, of course, for several years. There were women there in greater numbers than there ever were in Thule (Danes, Canadians, Newfoundlanders, British, and a few American wives).

One of our surveyors, Bill Stead, left us there and went to work as a civilian and brought his wife up. Some of our people were strangers, short timers, like the pipeliners who only stop for a few weeks. Perhaps because some of the Danish knew some of our people from Thule and BW8, there was some familiarity with these people. Whatever the reason, the atmosphere was strained at the base on this trip. I remember the club officer being particularly disturbed.

While we were at Ikateq, we waited for a helicopter part to be flown in from Narsarssuak. It had just snowed five inches or so, and there was no outline of the strip, so I went ashore and located the ends of the gravel strip and centerline. I marked the strip by walking short steps along each end of the strip and then walking straight down the centerline. This gave the pilot delivering the part a good target to aim at.

As I said, there was a tall mountain to the north and takeoff to the north. Traffic should land from the south and take off the same way. I had just finished my foot marks along the strip and was at the north end when the SA16 (amphibian) came in and landed almost at my feet, splashing snow and mud all over. When the pilot got out, he was the aforementioned club officer at BW1. He was happy and grateful to be on the ground and shaking my hand warmly.

Later, when we stopped in at Narsarssuak on our way, two months later, to St. John's, he was tremendously cordial. This was even before I threw a party at the club and spent the money I had made in the nightly poker game in the ward room on the *Westwind*.

In the meantime, we continued to design the ice cap stations on the DEW Line Extension. There was much talk of setting up sled trains to haul the materials, several thousand tons, from BW8 to the two ice cap sites. This was a badly crevassed area; we had finally gotten a road built up on the ice cap at Thule and filled some crevasses, using gravel fill and so forth, and we were already hauling swings.

A swing is a group of several tractors hauling sleds with Wanigans on them, plus cargo sleds. Wanigans were wooden and tar paper structures on top of sleds. These we used for housing, messing, and administration. The sleeping Wanigans had two shelves, like railroad berths, except that three men could sleep on each shelf.

I remember Paul Victor asking me how I liked the swing I took out to Site 2, where later we built Camp Century. At that time, it was a research station only because the radar placed in 1953 was obsolete.

The Wanigans were heated by wood-burning stoves. For safety, there was a hatch in the roof of each Wanigan. This had been made by cutting a hole in the roof, which entailed cutting into some fiberglass insulation and supposedly framing the opening with 2x4s. The frame was left out in some of the Wanigans, so when the wind was in the right quarter and the sled train was proceeding into the wind, smoke would accumulate in the Wanigan, so one would kick the hatch open to let the smoke out. This, of course, would cause bits of fiberglass to drop on the occupants and get down your neck. As I told Paul, you had a choice of getting your eyeballs smoked or enduring irritants caused by fiberglass in your skin. This was gradually changed, and things were happier on the trains and swings after that.

The alternate to the sled trains at BW8, with all the tractors and men required, was to use ski planes. Delays of a year of two because of crevasse and melting conditions ensued from hundreds of tractors we did not use. The beauties of air power with proper ski-equipped airplanes were what we wanted.

We had, at that time of course, some C47s on skis, but they could not haul much. We had some Fairchild C123s, which were satisfactory as ski planes but were small and couldn't haul pieces longer than eighteen feet and heavier than seven tons. Lockheed had, by this time, developed C130s, which could haul cargo up to 8′x10′ in size and width and height about 40′ long and twenty-five tons in weight. It could take off and land on very short, rough runways. With skis, it might have to have less weight but should be able to do the job.

One of my classmates at MIT, Chuck Hadley, was then Vice President of Military Sales at Lockheed. I wrote to him; he had been my sophomore right tackle when I was a sophomore right guard. He sent me photos and specs on the C130. We talked to the Corps of Engineers and the Air Force, and the race was on.

The first test of a ski-equipped C130 was at Lake Bemidjie, Minnesota. It was a failure, and Lockheed went back to the drawing board. Then, a C130 was turned out and passed the test at Lake Bemidjie, so they sent one to Sondrestrom. This one landed on the ice cap, but on takeoff, it hit a hard piece of sastrugi and tore off the nose ski and nose wheel. (Sastrugi is the Russian word for "snow drift.") When they get packed by the wind, rising temperatures, and pressure, they get hard, sometimes like a rasp.

Anyway, they brought the C130 back to BW8. The runway was well coated with foam. Everyone in the airplane except the pilot got into the tail area, and the pilot landed it with sparks flying and flying foam. A low-bed trailer was placed under the nose, and the airplane rested on the edge of the taxiway like a wounded duck.

In the meantime, my neck and the necks of many other people were really stuck out a mile. We had designed the stations on the basis of this airplane, and BW8 was stacked with steel beams. Then the word came through that this was it, and Col. Turk, USAF, a real pro, arrived with two C130s on skis.

I collected my partners, Hosmer, Philips, and Moore, and the four of us were to go out with the first flight and lay out the stations. It was 0°F at BW8 when we took off. Normally, the temperature gradient would be that the station at 8,000 feet would be -40°F, but there was an inversion, and it was a comfortable 0°F when we got there. We laid out the stations and left.

Our site people and the contractor's people stayed there to start construction, but they couldn't get out of the huts for several days, because it got to be about -40°F and blowing snow. But from then on, construction was okay. It is interesting to note that the work at both stations was done under a hard-money, low-bid contract. There was a clause about extra payment for inclement weather, but it never had to be invoked.

We used the C123 to set up camps on the ice cap for these stations. The coordinates for these sites were classified. I never knew why, because commercial airlines flew over the icecap and could have spotted the locations immediately. However, I went to New York City and was given the coordinates of each station, which I was to carry, unclassified, in my diary; like $73.15 or $16.35.

Then Bill Moore, my chief surveyor, and a crew of us landed at the appropriate location for the site after taking some sunshots. We had to move a couple of times, until we were at the more correct location. Then, the contractor set up a camp of Jamesway huts, and Western Electric sent in two expert navigators, Dr. Joe Chamberlain and Dr. Tom Nicholson.

They fine-tuned navigation and located the exact points the electronic people wanted, then hell broke loose. Maj. Buzz Commons, who commanded the detachment of C123s and also some C54s, called me at Thule to say we must change the coordinates forty miles south and forty miles west on one of the stations. So, we changed them and moved the sites.

Tom and Joe had just finished this work when Western Electric sent in their electronic people to make path loss tests to check the tropospheric scatter use of the site. The path loss tests showed that one of the stations was in the wrong place, and the signals were bouncing off a mountain some place.

I went back again. This time, we found out the coordinates I had been given were okay, and the change of forty miles south, etc. had been made on some other coordinates I had never heard of, the year before. Therefore, the original coordinates I had were okay. This was definitely an error on the part of the Air Force and Western Electric. If they had not been so security stupid, they could have sent the message in code, instead of forty miles south and so forth, and we would not have gotten into that bind.

I thought it would end there except, later, when all was going well and the stations were being built, an Air Force colonel at BW8, a new commander, gave a briefing for Gen. Wilson, the Chief of Engineers, Gen. Renshaw, Division Engineer, and several high-class civilians, Air Force generals, etc. One of his turncards listed coordinates as M&E's mistake. When he started to talk about this, I got up and said, "Colonel, that is not true."

When I was walking out of the place with Gen. Wilson and Gen. Renshaw, Renshaw asked me if I was sure of my facts. I said of course, and it could stand investigation. Col. Gould finally got relieved of command at BW8 and was sent to Thule as a supply officer to serve out the rest of his Arctic year. He had the moxie to come to my office in Thule later and ask my help in some matter. I don't remember what it was, but I did him a favor.

He also gave me a bad time in BW8 before he was removed. Western Electric had hired my old friend, Adm. Macmillan, the old Arctic explorer, to brief people on living on the ice cap. Old Mac had been with Peary partway to the pole and had sailed many times in the Arctic on his schooner, *Bowdoin,* but he had never been on the ice cap. He had always wanted to go there. However, the medicos at BW8 (he was in his eighties) would not allow his visiting these stations, because they were around 8,000 feet in elevation. I was going on a routine visit to Thule. Mac was going, and so was Count Eigil Knuth.

The count was a Danish explorer who had traveled around more of Pearyland than old Peary ever did. We were going up with Buzz Commons C54, and we told Col. Gould that we did not want a big VIP reception, because the Danes would take care of Knuth, my people would meet me, and I would take the admiral out to the ice cap on the Thule takeoff, so he would see the ice cap and walk on it.

Gould screwed us up. When we got to Thule, the limos with flags and all that met us and my people picked me up, but Mac and Eigel got mixed up with receptions and so forth, and Mac never got to see the ice cap that time. The irony of his not being allowed to visit the ice cap site at 8,000 feet but being allowed on an unpressurized plane at 12,000 to 13,000 feet on the way to Thule is the kind of irony of which the world is full.

The C123 story does not end here. These machines did yeoman work for us in setting up these stations. This was in 1958, and in 1981, I was at a meeting in Rome at the Comité de l'Arctique, which is the Arctic group from all over the world. I had been invited by Prof. Louie Rey who is a professor at a school in Switzerland and very interested in Arctic affairs. He wrote a book about the Arctic and included a photograph of our Arctic DEW Line site, Dye 3. The Smithsonian also showed pictures of this Dye 3 and only mentioned CRREL, no mention of the Corps of Engineers nor M&E, but we were used to that.

The conference was very high class, held in the chancellery of the Vatican. We had access to the Vatican Library and seats at the first audience the Pope had given since he was shot earlier. My Maggie and I flew out from Cairo to attend. There were many functions, ambassadors, and so forth. The reason I mention it here is that my old ice cap friend, Tom Nicholson, was at the meeting, and he was happy to tell Prof. Rey that the last time we met was in the middle of the Greenland ice cap in 1958. He was still boiling at Col. Gould's action back in 1958 (this is going back from 1981 to 1958).

The C123s were made by Fairchild. Their tech rep. on this mission was a man named Roche, whose sister lived in Swampscott and was a great friend of my sister in high school. The C130s went on to the bigger and better things in Antarctica, and for one thing, the Army, Navy, and Air Force all call the aircraft C130 without all the profusion of names given to most other airplanes used by several services. We got involved in several other incidents that could be said to be a bit outside our basic mission.

Nord was a Danish name and a U.S. name of a station for radio and weather built on the northern part of Greenland above the ice cap.

In the days of the building of Nord, a British explorer, Commander Simpson, had set up a station on the Greenland ice cap at about 8,000-foot elevation. He and his people had come in from the east coast and traveled over land and glacial snow to what they called "British North Greenland Expedition." The Royal Air Force, with the cooperation of the USAF, had set up an airlift from Thule. Materials and equipment were shipped from Thule by sea, and a squadron of RAF Hastings (bombers, adapted for cargo) was shipping these supplies to Commander Simpson.

This group was led by an Irishman named Clancy. The USAF had a young captain, Smoker Stover, operating their end of the mission. Now, the Greenland ice cap looks level, but in this particular area, it is sloped toward the southeast at about three-quarters to one percent.

Clancy always tried to make his drops as close to the camp as possible and as close to the surface as he could. The radio altimeter, however, does not read the distance to the top of the snow; it reads it about 180 feet down, about the elevation of the relatively solid ice. On one of Clancy's flights, he was in level flight and dropping 55-gallon drums of fuel for the camp stoves and for operation of the weasels. These weasels are tank-like vehicles, lightweight

Caterpillar tracks made of rubber, and had been used for years for traveling in snow and mud. Stover was with a drop crew of the RAF, kicking out the drums.

The aircraft flew right into the ice cap. Stover was thrown forward and crashed into the bulkhead and broke his back. Several of the RAF men were injured but not so seriously. It was 40° below zero. The airplane, except for two of the engines, was dropped off at impact in one piece. The airplane cabin was closed off as well as possible, and a heater installed. It was still not above freezing in the enclosure. Stover said afterward that he hated to urinate and lose all that heat.

At that time, Capt. Burnett of the USAF was near Thule with his C47 with skis, but some political operators wished to get publicity for Grumman, so it was decided to take a Grumman SA16 Albatross, which was equipped with a single ski on its belly. This use would make it a triphibian instead of amphibian. (I never could agree with the lexicography of that term.) I was at BW8 or BW1 at the time and knew an excellent pilot, Major Pierre Julin from Nebraska, who had many hours in SA16s, both in air and on water around southern Greenland. He and I flew up to Thule in an old B17. The SA16 was at Thule and so was Col. Balchen, who of course had participated in several rescues on the ice cap.

Bernt and I had been looking into a route across the ice cap so travel to Nord would not be confined to air. Julin told me he was a so-called expert in the SA16 at sea level but knew nothing about ice caps or snow. I told him I had been in ski planes and so forth, and at that elevation and temperature, the snow would be more like dry sand and not well lubricated. There would have to be some rocking of the airplane in getting it underway, before he could cut into his JATO.

In a day or so, the weather cleared, and Julin set out. I was in a C54 with Bernt Balchen. We were alone in the passenger compartment except for occasional visits by the navigator. I remember, on the way, Bernt said to me, "John, you engineers have got to find a road across the ice cap."

I said that it was possible, but the access to the ground at Germanialand to travel toward Nord was difficult. The ice cap at that point was called Cape George Cohn and had about a 500-foot ice cliff.

Bernt shrugged. "You just take a blade," and he made a sweeping gesture. Then he went back to reading a French novel in paperback. Of course, the most famous line ever quoted from Bernt was, "Ve do it."

We flew overhead as Julin made his landing. The landing skis cooled and then stopped near Clancy's airplane. The uninjured RAF and Simpson's people and the paramedics who were with Julin got Stover and an RAF sergeant into the SA16. Julin started up the airplane.

Bernt was talking on the radio, "Viggle the vings."

Between the power of the engines and the crowd "viggling the vings," the SA16 gradually picked up speed. Then, Bernt yelled, *"Now!"* and Julin kicked in his JATO. The SA16 flew back to Thule.

Later, one of the paramedics told me that he had kept a frozen foot of one of the RAF sergeants against his bare stomach all the way to Thule. These paramedics were double paras; they were also parachuters who had jumped several times to help downed fliers.

After three days of bad weather, the rest of the RAF, including Clancy, were picked up without incident by Capt. Burnett in the C47. When the press reported it, it described in detail the subject of the SA16 and Julin, but there was no mention of Burnett, just word that the rest of the group was later picked up by another aircraft. Someone gave me a copy of a book written about that time, *Rescue Below Zero,* that describes this whole affair, more or less.

Burnett was destined to be a part of another almost faux pas that fall. He was told he must pick up some people at Nord not earlier than a certain date. It should have read *not later than* that day. When he got to Nord, there was lots of snow on the ground and in the air. The airfield was not well marked. On takeoff, he hit an oil drum on the end of the strip and flew to Thule with a damaged aircraft. If he'd stopped, he might have been at Nord all winter.

There are two sequels to this résumé. First: Capt. Stover recovered and went back to flight duty, then his airplane hit a rock in a cloud, and he perished. Second: The USAF sent an SA16 up on the ice cap near BW1 for practice. They couldn't seem to get the plane off, so they sent a C47 ski plane up there. The group finally got the SA16 off but managed to leave the C47 behind. I believe this was the last time Grumman and the Air Force tried out the SA16 as a ski plane.

Burnett and I met several years later in New Zealand. He and Capt. Rinkle were flying C124s then and hauled us, me, and my party down to Antarctica in 1957. We met at the operations building at the airport in Christchurch, each saying to the other, "What are you doing here?"

Capt. Rinkle had landed and taken off from a compacted snow runway built on the Greenland ice cap in a fully-loaded C124 in 1953-54. Our Capt. John Napier, Corps of Engineers, later a civilian employee of M&E and Universal Engineering, had been in charge of the operation. Napier had used equipment similar to a harrows, which applied heat to the blades and, followed by rollers like a rototiller, compacted with snow with a little heat, just like a snowball. The heat and pressure from the hands compacted and hardened it.

Speaking of public relations, the Navy had a bit of publicity by building a compacted snow runway a few years later. Our man, Irv Call, was an old Navy man. He went out on one of the flights, and the plane almost got wrecked.

When the Navy came to Thule to do this project, the base commander introduced me to the Navy captain and a civilian physicist and told them I knew something about the ice cap and that they should talk to me. I asked them if they had read Wegener, Paul Victor, Henri Bader, Nagaya, and the Corps of Engineers' work on the previous runway that Napier had built. They told me politely that they had not, and they did not intend to look into what had been done before. They just wanted to do it on their own. And they did, with much less success than Napier.

There were already some Navy people at the site, and these two were going to join them. Major Welch, who was sort of Office Engineer for the Corps of Engineers, called me and asked if I would lend them a compass. I pointed out that we had compasses, but they were very confusing to say the least, and they should navigate by the sun and watch. But they wanted the compass. We had made tests there and found that there were several degrees variation per hour, up to 32 degrees per day.

Anyway, they set out on a weasel and traveled around in circles and finally found the camp. By this time, they were both snow blind and frostbitten. They had to keep getting out of the weasel so the metal would not interfere with the compass and take off their snow goggles. I suppose I should recite the old Explorers Club maxim: "To have an adventure is the mark of the non-professional."

In 1957-58, I managed to spend almost the whole year in the daylight. I came early to Greenland for spring work at Baffin Island and Greenland and sailed on the *Arctic Sealer* for three months in the fall of the 1957 and left for Antarctica right after that, for the Antarctic summer, and got back in Greenland in 1958 for the Arctic spring and summer.

In 1957, we continued our preparation for construction of the extension of the DEW Line across the ice cap and into Iceland. I prepared a proposal for a survey of three of the land sites that we had looked at on the *Westwind*. One at Qaqatoqak on the west coast; one at Kulusuk on the east coast; and one at Kangek near the southern tip of Greenland.

In my proposal, I set up the following. We would hire a Danish ship, *Maggadan*, or her sister, *Kistadan*, take a helicopter furnished by the USAF, and man the expedition with twenty-one people–surveyors, soils engineers, and a blaster to blast test pits. We would furnish all material for surveys, etc.—targets, poles, sounding equipment—plus all required labor and material and equipment, except for helicopters and explosives, which were to be furnished by the Air Force and the Army. I offered to do all this in ninety days for $340,000.

The Air Force considered this excessive and would not believe my estimates of travel time, weather delays, and actual survey time. They said they would hire a ship, and the labor and materials would be paid under a cost-plus fixed-fee arrangement. Ninety days proved to be a proper estimate, and between hiring the ship and paying for our people and material, it ended up costing the Air Force over $350,000. We and the Air Force military and civilian personnel did not have as pleasant a trip as we would have had on a better-equipped ship like *Maggadan*.

The Air Force chartered a ship from some people in Halifax. It was an old Navy deep-sea tug, about 1,250 tons. It had been built at Bath Iron Works in Maine. After WWII, it had been named the *Port of Beaumont* from Texas and was sailed by Capt. Finn Ronne of the US Navy Reserves to Antarctica. It was left frozen in the ice one winter and carried on extensive exploration in the Palmer peninsula in Antarctica. (That story is a tale by itself.)

One of the geologists on Ronne's expedition was Dr. Robert L. Nichols from Tufts University, who later worked for us, near us in the Arctic, and worked on my party in Antarctica reconnaissance work in the Marble Point area. The *Arctic Sealer* was a wooden ship about 190 feet long. She had two inches of extra greenheart timber to help out on the ice and had several years of work in the Arctic on resupply and the DEW Line in Canada. It was hardly a substitute for the cleanly Danish steel ship, but somehow, she got the job done.

We formed our group at Fort Pepperell in St. John's, Newfoundland. Most of our surveyors were veterans of our Arctic work. We did have a handful of rodmen, one new survey chief, and our blaster, who was to work with Magnusson, our soils engineer. There was some delay in St. John's in getting the helicopter deck built on the ship and more delay getting the Air Force radio installed. The *Maggadan* already had plenty of radio and plenty of helicopter deck.

We finally sailed. I wanted to sail first to the west coast and then to Kulusuk on the east coast, then to Kangek on the south coast. However, Capt. Lally, that brave man who volunteered to face another three months of seasickness, had talked with his people and thought we should go to Qaqatoqak last, because of its location on the west coast where the weather is better than the east coast. He'd even called Capt. Peter Smenton of the *Westwind* and got the same opinion.

Of course, neither one had ever been to Qaqatoqak and were not thinking of the peak, 5,000 feet at the edge of the ice cap. They were only thinking of the latitude at sea level on the west coast being warmer than the east coast. We kept up a continuing running battle over the sequence, and finally, I won out, and we at last went to Dye 1, Qaqatoqak second. Even then, we barely made it before the real heavy snow hit.

At Kulusuk, Dye 4, we anchored the *Sealer* in the passage between the fjord Toikateq and the island of Kulusuk and started our survey. I immediately started my personal reconnaissance on the survey at Kulusuk to make a runway to substitute for the old one at Ikateq. Lally, backed up by his command, fought against this and, as I said before, even invoked my friend Balchen's name, which he thought might be besmirched if we built the strip at Kulusuk.

We set up an emergency cache of food and supplies and clothes and a couple of tents on the shore, in case we lost the ship by fire or by ice. Magnusson and the blaster lived onshore, and each morning at 6:00 a.m. started the day's work with a blast by a 15- or 40-pound shape charge. These charges were in fine supply all over the Army's depots. I had of course used them in the Army as an engineer soldier and had, over the years, applied the principle in making mud caps. I had found in the prior years in the Arctic that the shape charges were very handy in blasting test pits in the frozen soil, so we had carried them on the *Westwind* and also on the *Sealer*.

We had a stock of these, as well as plastic explosives, C4; the idea was to use a higher standoff than comes with the charges and blast a hole in the ground. This hole may be six to eight feet deep and, with a shovel, like a pole digging shovel, can be excavated more. We cleaned out the loose dirt, got a hole eight-feet deep by a foot in diameter, and then filled this hole with C4 and fired it. This gives you a large hole in fine-grained permafrost that you can get down into and clean off the face, to get the stratigraphy and sample the soil for ice concentration and so forth.

After we had been at Kulusuk a few days, we got a radio message that a Norwegian Sealer was nipped in the ice some days sail north. We were the only ship in the area. The men had abandoned ship and were out on the ice. We left some of our people there, at Angmagssalik at the hospital, and a few of us were going with the helicopter crew.

I went ashore that night and blew our store of powder that we had on the beach. We did not want the Greenlanders on the island to get at the stuff. My caution was useless; they wouldn't touch it, anyway. With all the food and clothing in the cache, no one touched a thing.

The Norwegian ship had gotten out of the ice, and the crew was back on board and were gone. None of the Greenlanders had touched the cache. When we left, we offered these people, poor people, the food we had left there and had to hand each one something by hand. You couldn't just tear off the tarp and say, "Come and get it."

We had just about as much bad weather there as we had expected. One day, the fog was right on the deck of the ship. No survey could be done, of course, and the helicopters were grounded, so some of us went fishing. We took the launch, which was actually a regular old lobster boat; with Lally, Capt. Little, TC officer along to check the beaches, and a couple of electronics mavens, plus the engineers and mate, and the ship's doctor.

We jigged for cod; the waters were teeming with them. We pulled them up by the tail, by the eye, everything. Once, the mate, engineer, and I had done this before, and we soon found ourselves like fishing guides—taking hooks out of the fish, smashing them against the box, and slapping them into the box.

Cod this big are best de-hooked by putting your fingers in their eyes. It stuns them, and then you take out the hook. These fish were too big to fit in your hand any other way. The others, all these old soldiers, were too squeamish to handle a fish like that.

Then, we pulled alongside the ship, and although the mate, engineer, and I showed the others how to clean the fish, they sort of wandered off. It ended up with the mate and me cleaning all the cod. We finally got over 350 pounds, all cleaned and split, in wooden tubs on the deck of the ship. The temperature was about 30°F then.

I'm very partial to poached fish, especially cod, if I have it. I told the cook, "Tomorrow morning, I want some poached cod."

"Yes, sir," he said.

Up to that time, my breakfast was half a grapefruit and a kippered herring. Every day, the cook used to cook five kippers–one for me, one for the captain, one for my chief surveyor, Herb Yamane, and one for the cook. The last one, they always threw over the side. I longed for fresh fish. It was not to be.

There used to be a poker game each Saturday night at twin decks. The group that played were our oldest surveyors, except for me and Herb Yamane, the ship's doctor, and sometimes Captain Miller. The night we left the fish on the deck, the mate, who was not allowed by the captain to have any rum, was, as usual, trying to cadge drinks from the players.

The mate, just trying to make conversation, asked, "What will you do with all those fish?"

Captain Miller, who was not exactly sober and no fish eater anyway, growled, "Throw 'em over the side."

"Yes, sir," said the mate, and he threw them over the side.

The next morning, not knowing about this, I sat down in the ward room early. The messboy, a Newfoundlander, put before me the same kipper. He did not know anything about the cod.

He went to the galley, and the cook told me there were no fish on the deck. There wasn't much I could do to the mate, but I did make him miserable with his hangover.

I told him to get the launch alongside with a can of extra gasoline and a deck hand and to take me through the pass or gut against a six-knot current, halfway around the island, to where there was a narrow gorge where I had landed a year before from the *Westwind*. I could look at some gravel deposits and then work my way about six miles or so along the road, to where the ship was anchored, and check out surveyors on the way.

I almost played it too hard. When he left me off on the rocks, he said, "This is a hell of a place to leave a man." He was really suffering with hangover. I told him I had been there before and that he should go back to the ship. When I appeared on the beach, he should come pick me up.

There must have been a higher tide or something that day, because just minutes after the little boat cleared that little bay, a large iceberg that had apparently been grounded for a long time moved across the end of the bay and closed it. If the boat were trapped in there, I was not sure we had enough powder on the ship to blow the berg up. But the mate was a fine seaman, and he and his crew were some of the best small-boat handlers I ever saw. Better than the Navy, and even almost better than the Coast Guard.

The ship had no evaporators, and the fresh water tanks were small. When the water was short, only the cooks were allowed to bathe. Not that they always took advantage of the chance. To get fresh water, we had to put in at a port where it was available or lay alongside an iceberg, which had a pool of water in it that could be pumped aboard. Sometimes, we would find a waterfall along a fjord where we could beach the ship and transmit water from the heights and into the ship with drums and hoses. We only lay alongside an iceberg once. It is chancy; if you've ever seen icebergs roll over and spitting and, now and then, roar like thunder, you'd think twice about getting next to an iceberg.

At Kulusuk one night, when we were at anchor and with almost a full gale blowing, we worried about moving the ship, because plenty of bergy bits were hitting us. I was on the quarter deck when an Inuit came along in a kayak. He swung his paddle and, in the crack, there was a note. As soon as I took it, he saluted and paddled away.

The note said that one of the chief men at the village had fallen and cracked his head. The midwife had sewn it up, but it evidently had not been cleaned properly, and the man had lots of head pain.

The doctor was all for leaving right away. The captain and I said it was too rough to put a boat over the side, and the rocks and surf made it impossible to get to the village, that we should wait until morning.

"I'll row you, Doc," said one of our surveyors. Anyhow no one left the ship, and at daylight, the engineer, Mr. James, and the mate, Mr. Drodge, took the doctor ashore. Mr. James said that the doctor performed proper surgery on the man's scalp and gave the Inuit enough medicines and full examinations to last him a long time.

Speaking of the "I'll row you," we made port later, and the eager beaver of that night and the doctor took a dory and went rowing. The crew was very amused. They couldn't do anything at all about it. All the captain could do was laugh and hope that they got back to the ship again. We were pleased that we didn't let them go out at night.

We had to stop in Narsarssuak to pick up a couple of Danes. One was an electronic type and the other, a construction man. On our way back down the fjord, the electric steering got out of whack, and we made a sharp right turn. Fortunately, it was a right turn, because the LSD Plymouth Rock was on our port side, passing us with good speed. If our ship had turned left, we would have been cut in half. Plymouth Rock is a small type of landing ship dock, but they still outweighed us by plenty.

This steering problem stayed away from us for most of the voyage. However, we were running away from a full gale one night, and we were practically on our way to Accra or someplace in Africa. Capt. Lally and I were sitting in the ward room, and not only had he accused me of being callous; he was talking of abandoning ship.

I said I had a bottle of Barbados rum, 151 proof, British Navy-style issue, and I was going to stay with the ship and go down happy. He left to go to his bed across the companionway.

Then, it seemed as if the ship had directly hit on an iceberg, because it shuddered and almost keeled over. The steering had gone bad again. We had made 90-degrees right rudder, 90 degrees again, and had made 180-degree turn. We were going north toward Greenland again, instead of south toward Africa. It took a while to get on proper course again, but I think we were lucky we didn't capsize.

On our way to Dye 1 from Kulusuk, we paid a short visit to Angmagssalik to send a radio message. We also had promised that the cook and one of the deck hands who were good guitar players and singers could perform for the patients in the hospital.

We didn't wish to put any boats over the side this time, so I hailed an Inuit in a dory, and he rowed me ashore. I think he asked for the equivalent of about fifteen cents or so. It was perhaps 9:00 p.m., and the sun shone on the little ripples in the bay, making them sparkle. The bay was bordered by rocky cliffs, and ice cap mountains were in the background.

On the way back, the cook and the deck hand were sitting on the stern of a larger dory. Two Inuit were pulling manfully on the oars. The guitars were playing, the boat was dancing over the little sparkling wavelets, and the sounds of Leadbelly's song, "Good Night Irene," floated over the little harbor. It was beautiful. That combination–the ice-capped hills, the sparkling water, and plaintive song–I have always remembered.

Dye 1, Qatatogak, as I said, was on a 5,000-foot mountain next to the ice cap almost exactly on the Arctic Circle. There was a small beach at the foot and a deep lake from which a small stream flows over boulders into the fjord. The fjord means "a place where there are many trout and ekalukgarsuit." The lake is full of salmon trout.

The Inuit fish for them by building weirs and traps, and though they seem to take a lot of fish, there are not many Inuit there. They come up from Ivislak, which is a small fishing port on the coast where the fjord meets the sea.

Near the lake, there are some graves of old Inuit made of stones piled closely, but with some chinks where you can look in and see the dead inside. These openings are not made so you can see inside, but so the souls can move in and out.

We looked in one of these, the *Arctic Sealer* doctor and I. The bodies were mummified, freeze-dried. I told the doctor which one was the woman.

He said, "How do you know? You can't see the pelvis..."

I said I could tell by how the teeth were worn. In that skull, the teeth were really worn down, and the other one wasn't. "That's the woman. The wife's teeth get worn from chewing seal skin to make boots and so forth."

Then, he became his own crazy self again. "I want to take at least one of these back to the U.S. with me for science."

I said, "Doctor, if you stir up this grave, take up even one stone, I'll kill you."

At first, he did not believe me. I repeated it again. We went away, and we never spoke to each other after that.

The last time I saw him was at the Officer's Club in Pepperell. He was walking on his hands across the dance floor. They threw him out of the club, and I saw him no more. Capt. Little was always amused at the doctor and wondered how he became an MD and a paratrooper.

I have a framed, sketched drawing by Captain Sorenson of the motor vessel, *Tikerak*. He was a Dane engaged in moving Greenlander cargo around the island. He was called "Super Sorenson," because he was a super—if not exactly superman—and quite a fellow.

The first time I met him, he was in the harbor at Julianehab [now Qaqortoq], and we came in with the *Westwind*. Captain Smenton of the *Westwind* knew him, and we exchanged visits. His ship was small, less than 180 feet. It was a wooden ship heavily built like the *Arctic Sealer* for working in ice. She had a fine interior of all polished mahogany and brass.

His personal girl stewardess was a very pretty girl. Captain Smenton had always joked that he would like to exchange his Filipino steward for the "Super's girl," but of course it never happened.

Super Sorenson spent some time telling a tale of Arsuk Fjord, which is in the Ivigtut, where the cryolite mine is and where, during WWII, the U.S. had a couple of batteries of artillery. Their commanding interest was to protect the traffic in Cryolite.

Sorenson made a sketch of it by hand, and I have a good navigational chart, in case I choose to sail again into Arsuk Fjord. All they do there now, I believe, is work on the stockpile; the ore is worked out. Cryolite is ice stone. It is so transparent, if you suspend it in a beaker of water, it is hard to see the outline. All you can see is the string. Anyway, it was used as anchors and was always a phenomenon when it sank in the water and became invisible. Cryolite is used as a catalyst in the electric furnace process of refining aluminum. The aluminum is dissolved in the cryolite. A great amount is now made synthetically.

Paul-Émile Victor was a French scientist, an explorer, and ski-trooper soldier. I have a book of his inscribed by Paul in Greenland. He had a camp called "Eismitte," after Wegener's operations, and had considerable reconnaissance and research on the ice cap. He called his place "Inland Ice." He has dual citizenship, French and American. He was a major in the United States Army and helped train the Tenth Mountain Division in Colorado before it went to Italy—skis and all.

When I first met Paul in 1952, we were planning a couple of AC&W stations on the ice cap, about a couple of hundred miles out of Thule. Paul had the idea of making them like submarines. The weight of the station would equal the weight of the displaced snow, to let the station sink with the ice cap as the accumulation/consolidation of snow beneath. We did not go quite as far with the concept as he'd wished. He figured we should have crew quarters where the bunks could be folded up into the building in the space used for work and duty.

We built these two stations completely by air power, using two C47 ski planes for personnel and electronic equipment and for air dropping structural items. While we were doing this, we were also laying out and building a ramp from Thule to the ice cap, to get the only crevasse known so that the tracks and the sled trains could travel to the sites and to other places as well. We used dog teams, Inuit driven; helicopters; sea planes; and just "feet" to do this with.

We had planned to fill some crevasses immediately but avoid them, if possible. I remember flying up with a Transportation Corp pilot in a helicopter, and only one person could be carried on this little Bell. We landed, and I started probing the snow for snow bridges. The pilot had to move his battery when he had no passengers for ballast. When he stepped away from the chopper, he almost fell into the black hole of a crevasse. After that, he never took his hand off the machine when he was in the snow.

Paul was working with the Transportation Corps in developing sled trains and routes. The D7 tractors pulled several sleds. Some were Wanigans, just houses on skids, bunk houses, two shelves sleeping three men each, and there was a wood-burning stove in each. Wood was safer than oil, and a few sleds could haul wood; each "swing" as the train was called. This was their first use in Greenland, although they had been in use a long time in Alaska and Canada.

Later, when the ramp at Thule takeoff was stabilized, we were working on the nuclear reactor, which was sited near the easterly AC&W site. We had regular store-bought sleds with housing, like RVs, in them. Paul was a few years later at Sondrestrom, but we never tried the sled trains there. We did it with the C130 on skis, and that method got us done down there and with the DEW Line Extension sites.

There were several scientists up there besides Paul. Dr. Henri Bader, a Swiss whose doctorate dissertation was on

"Metamorphosis of Snow." He taught us a lot. Not as much, though, as Nagaya, a Japanese scientist who had done much work on snow physics with the snow on overhead cables and pipes in the mountains of Japan. We learned to be appreciative of this problem in our work in the snow.

Bader questioned Paul's doctorate. "Is he a doctor? I don't think so." However, I later found in "Who's Who in Science" that Victor was listed, Father Dan was listed, but Bader is not, so I don't know.

It was while we were engaged in the nuclear plant that I saw Carl Eklund for the last time. We were in the same tent, and he was trying to find the difference between the Arctic tern and the Antarctic tern. I guess they were the same bird, but they fly back and forth. I lent him a pair of tweezers so he could clean out a specimen. Anyway, he got sick and had to be flown out, and I never saw him again.

We had some real characters as photographers in the Arctic and Antarctic. At Thule, there were Doug Downs, Capt. Rosenbaum, Otto Hess, and Allesandro. Downs was not there as long and was a genuine operator; he rose to the occasion. One time, there was no film at all, and the VIPs were due in. However, Downs got all his men and his cameras together and flash bulbs, and they went out and shot the planes. All were happy, all looked forward to the next day, and, of course, no one ever got any pictures.

Rosenbaum was practicing with what I suppose you could call a "spy" camera. He took my picture (my wife has a copy) on a negative about the size of the head of a pin, or so it seemed, and he blew it up to a 4" x 5" size. He entered it in a contest; he had me sign a release. But results were zilch; he didn't win.

Allesandro was a real "skirt chaser." He could always get the new girls undressed by showing them glamorous bare photos that he had taken of other girls, promising what he could do for each one, if she wished to show up. In gratitude, the girls would often go to bed with him, I guess.

Otto Hess was the best, an excellent photographer but grouchy, hard to get along with, and a real drunk. We, of course, always built the field darkroom with less than five feet in any dimension, so the photographers couldn't turn on the red light and go to sleep.

Otto finally got fired. He was asleep in his room at 2:00 or 3:00 a.m., and there was a bad accident on the road to Thule. He was ordered to go out and take pictures. He drove out and took what turned out to be excellent photos.

On his return to quarters, he ran right into the building—he was drunk. The Corps of Engineers discharged him for being drunk on duty.

Otto sued and won, getting back pay and damages. His defense was that they knew he would be drunk at that time—he was not on company time. They should have provided him with a chauffeur, if they wanted to use him at that time. He won the case.

Inuit—*human beings*. Until the Europeans came, they believed they were the only human beings on Earth.

They are spread thinly from Siberia across Alaska, across Canada, and all across Greenland to the eastern shores. They are not many; the whole group could not fill one of our fair-sized stadiums. And they are dying out or being assimilated into other cultures.

I met them in Greenland and Northwest Territories of Canada. The language they speak was not written. They can understand one another all across the Arctic Circle. They live above the tree line; the woods are not for them. They wished that English was taught in their schools instead of Danish, because Danish is harder to learn.

Life is hard, yet they contradict the scientists who say that it is not in the genes of Homo sapiens to be aggressive and make war and be violent, and it is said this is true, that aggression is caused by the leaders forcing their subjects to war, for example. I really don't know where the leaders get their genes; perhaps the old divine right of kings was a true statement and people who are born to the purple and whose genes make them leaders or the people in wars.

Evidently, there are none of these leaders among the Inuit, because they do not even understand what war is and how people can form armies and group together to fight other nations. It is said by some that violence is born amongst groups who are deprived and forced to live in terrible surroundings. This may be true of the people who live in slums, but with the Inuit who live in hard conditions, very hard conditions, they are non-violent toward other people.

They do not steal; they respect caches; they have no political bosses; they respect the best hunters but live in family groups, and no orders are even given. They cannot lie, and they will not give their word lightly. If you ask them to go hunting tomorrow, they will only say "maybe," because they know it may be bad weather; if there is a white-out, hunting would be impossible. So, if the weather is okay later, they will hitch up the dogs and off they go.

As de Poncins said in *Kabloona,* when he met the Inuit at Pelly Bay, "There they were, smiling, shining eyes, the true Inuit—the people who could not lie."

Of course, in Greenland now, the Greenlanders have changed the names of the towns, and Gothaab, for example, the Danish capital is now Nuuk. It will be interesting to see how the political change will change these people.

I think it was Hemingway who once said, "War is the health of the state." Perhaps now that there is a state where before there was none, Greenlanders may become warlike.

The Danes, as badly as they have treated the Inuit, cheated them in trading foods for furs and ivory, brought them whiskey, tobacco, sugar, and Coca Cola, I believe they have done better than the Canadians or than the Americans. The Canadians, the Christian missionaries, set it off in a sort of hieroglyphics to express the Inuit language.

The Danes tried to translate the old oral words into Roman characters, so the word for, say, "the place where there are many trout" became eskarukgarsuit and was written that way. Then, they taught the school children in Greenland as a first language and

Danish as a second language, so there was no culture block, and the children could talk with their parents in the same language. This method appealed to me. It was right for Greenland. I do not agree we should do the same thing in reverse, however, where the U.S. teaches Spanish children in Spanish as a first language and English as a second. This is the USA, and Greenland is Greenland.

I have heard Greenlanders say they wished that English was taught in their schools instead of Danish, because Danish is harder to learn and is really not spoken as much around the world—only 5 million versus about 500 million.

Inuit are very quick to learn on how to operate construction equipment, the use of tools, rifles, radios, etc. Their preventative maintenance leaves something to be desired, but of course, maintenance in the U.S. is also a sometimes thing.

In the 1950s, I met an old Inuit in Northern Canada, just across the bay from Greenland. He spoke some English, and I knew a few words of Inuit. He told me that the Inuit could sail big ships and fly big airplanes if they wished, which I am sure they could learn to do; but this was no way to get seals.

He said that the white man brings rifles and ammunition and power boats, good for getting seals. And then he said, "You know the white man has three gods, but they say they only have one. But the Inuit has one big spirit and, of course, a crew of animalistic spirits, and they believe that the soul doesn't go to heaven. It just stays somewhere near the people, but it is invisible."

I disagreed about the three gods but, when thinking about it, remembered that the area where we were talking had Anglican, Catholic, and Moravian missionaries and each one taught *This is the word of God,* the true words.

I have not been in the Arctic since 1963, but I went to a conference of the Comité de l'Arctique-Louis Rey, in Rome. One of the participants was a Catholic priest from Pond Invet, near where the old Inuit and I met.

Pere Guy was a slim athletic looking man, in contrast to the old Pere Daniel, an old bearded fellow who had been there years ago.

This Pere Guy was an explorer and scientist and fairly ecumenical. I asked him who cared for his flock while he was away, and he answered the Anglican minister took care of them while he was away, and he reciprocated when the minister traveled. He said that was okay, and the Inuit enjoyed it. He thought they got a benefit either way.

There was a German woman there at the conference. She and her husband were ice scientists and operated a laboratory at Pond Inlet. When I told her what the old Inuit had said twenty years before, she said he must have been really old—the natives were not like that at all anymore. In fact, one of them had just married one of her daughters, and they were training him to work their laboratory.

I found out that the Inuit never beat their children. They train them by kind words, sometimes with a little teasing or lampooning. Inuit children are never disciplined and nearly always allowed to do what they please. They are amazingly well behaved.

Their education is based the principle that a child given its own way will learn by practical experience what is bad and will without much coercion choose the good. As a result, Inuit children are usually very independent and full of self-confidence.

In summer especially, they play half the night. It would never occur to the adults to send them to bed at a specific time. They sleep when they are tired and eat when they are hungry. Just as adults in a group in former days were kept in line through the rule of public opinion and transgressions, where transgressions would curb with lampoon, songs, etc., some of the young control one another by an occasional dose of ridicule.

Inuit parents do not approve of striking a child because then the child will feel ugly, and he will want to strike someone else. They imitate their parents' activities in play. Small boys hunt birds with bows and arrows, while girls mother dolls made of skis and carry them in their arms. Boy and girls learn early to do manly things or womanly things.

The Inuit live in probably the harshest and most forbidding part of the world. Short summers with plenty of mosquitoes, except in the Thule area where they aren't any bugs. It's a murderous climate, with the continuing threat of famine, but they are gregarious, good, and amazingly contented. Optimism stays with them in adversity.

"The caribou will come," he will say. And the days pass and the caribou don't come. They just say, "Can't be helped." Just like you say in Jamaica after some minor accident.

There, I heard one of the people say that you can't help: "It's not your fault, the road's too narrow." When we were in Jamaica, the people were rough about a lot of things, but they were polite in the automobiles, giving way to someone else.

To the Inuit, as a rule of public opinion, ostracism is their worst punishment. Their approval is a man's highest reward. Their civilization is a savage civilization that has been aborted. They look after their parents, and they honor the aged, but in a famine the old people have to die. They accept this—get rid of a life that is no longer a pleasure, and also to relieve the family of trouble.

A hunter's pleasure is to share his skill and good fortune. His sole reward is the approbation of his equals. They were masters of their fate, since they knew no other. They were alone in this best of all possible worlds. They were happy and content.

Somewhere, the white man's culture destroyed the fragile fabric of this ancient way of life. The concepts of master and servant, of material wealth as the measure of a man's worth, the subordination of one man to another—all this was alien and mystifying to the Inuit. Then, the white man destroyed much of the Arctic's wildlife, which is the basis of the Inuit existence. They left the Inuit with a legacy of mass starvation, disease, and dependence on the white man.

Measles wiped so many of them out. We were quarantined in Goose Bay and not allowed to go to Greenland, where they would infect a lot of the natives there. Now, most of the Inuit are in a high-tech state, and most of them linked to their ancient hunting culture, which is now dead, and not yet part of the white man's. The Arctic world and its material compensation and the status symbols of a stratified society for which the young in particular yearn for the white man's world.

There was an Icelandic Sea captain at Kulusuk when we were there, and he beat up a girl who wanted to be paid whatever it was, or she didn't want and he tried to rape her, or he did and she wanted money for it. He beat the girl up, and the young men in the group there pretty near killed him. I remember he had false teeth, and they broke his false teeth, so he almost choked. But later, they would come and say, "Greenland is ours," at Kangek in 1959.

In some of the work we had over there in some of the southern areas, there are dangers at night now, because the Inuit and the Greenlanders don't like to have people around there who build any more of these arctic constructions by the military. They are brave people, though.

When I was in at Resolute Bay a few years ago, an Inuit was in the hospital there, lying on his stomach all summer, because he had been hunting a polar bear with a 22 rifle. There were two of them, and each had a 22 rifle, and they had two teams of dogs.

He got close to the polar bear, the dogs got loose, and they jumped at the bear. The Inuit got close and shot right through the heart. With a .22 rifle, he only had a chance at one shot, and this time the man missed and the polar bear was charging. The dogs, like good hunting dogs, couldn't believe he'd missed, so they were running off after the other bear. By the time his buddy got there and shot the bear, the hunter had gotten his behind clawed a little bit.

Those dog teams can haul a loaded sled of about 800 pounds and go forty miles a day. Sometimes, they can go faster than that and make 100 miles in twenty-four hours. Peary coming back from the Pole made seventy miles a day and that was just 3-1/2 miles an hour for twenty hours. It's quite an art, and in fact, in the exploration in the Arctic now, I still think that dog teams and ski planes are better than tractors and snowmobiles. The Inuit have these snow skidoos and snowmobiles. Only in Thule do they still have dog teams, and they don't use them up there.

The Inuit are very hospitable, have charming manners, and they say hello in a bunch of different ways to greet people, because they never say goodbye. When somebody is going to leave, they just leave. Nobody says goodbye.

In 1850, a ship came in there, and the old Inuit said, "The ship was full of iron and everything that was precious in our land." That was in 1850, and it reminded me of what Herodotus said before 450 B.C.: "It was so then that iron had been discovered to the hurt of man."

Almost always, when I talked and showed slides of the Arctic and Antarctic, with pictures of Inuit, someone would always ask me if the men would offer their wives to visitors. I could only say I never found this. When the girls are young, they may be like other primitive tribes and seemingly promiscuous, and if they become pregnant and give birth, they become more valuable as wives. Of course, there was no illegitimacy.

However, I found there were times when a hunter needed to get his seals for the year and his wife was pregnant and not fit for travel, he would borrow a wife of another hunter who already had his seals. A woman is essential on a hunt.

The hunter builds a snow house and drives the dogs and hunts. He has no time to cook, keep the lamp trim, and keep the clothes in shape. He hunts, and he needs a woman to keep his house for him. When the white man hunts, he has no woman to do this; the men do all these chores by themselves, although poorly, making the hunt inefficient. Of course, if the hunter has another man's wife in his igloo, sex may happen, and if so, it is okay.

The people in the Arctic poor areas are changing, and I trust that a new warlike Spartan state will not emerge. But it seems that when the two cultures mix, it is not always possible for the best of each culture to survive. In 1957, I heard some young Greenlander shouting, "Greenland is ours." And in Barbados ten years later, I heard some young Blacks shouting, "Black is beautiful."

On one fall day in 1957, we could not survey; the fog was low on the deck of the ship, so we were in the launch, fishing for cod. A hunter in his kayak came alongside. He had a rifle, a harpoon, and a float made of seal bladder latched to his kayak. I asked him in my stumbling Inuit if he wanted a fish.

When I heaved a cod of about thirty pounds onto his deck, he smiled and said in English, "Thank you very much," and then paddled away.

In fact, you might as well say that the "iron had been discovered to the hurt of man" here. At any rate let me say that the Inuit welcomed their visitors and were hospitable, but when the visitors leave, there are no farewells. They just pack up and go. I think that perhaps the Inuit have many words for welcome and for hello but no words for goodbye.

♦ ♦ ♦

*JJS, Greenland*

*SW Coast near Blule West One (Narsavsag)*

*Scraper Prepared for Aerial Delivery*

*D-4 Prepared for Aerial Delivery, 1953*

*Lima M34 Prepared for Aerial Delivery*

*Cat Prepared for Aerial Delivery, 1953*

*Dew East Facility*

*DYE 1*

*BMEWS*

*P Mountain*

*BMEWS 2*

*BW1 1952 Survey 1*

*Angmasgssalik 1*

*Angmagssalik 2*

*Angmasgssalik 3*

*LCDR Stephens & JJS ascending the Ice Cap*

*Kulusuk*

*Kulusuk Aerial*

*Kulusuk Group*

*Greenland Stations*

*Camp Century*

*Edge of Ice Cap BW8*

*Ikateq 1*

*Ikateq 2*

*Permafrost Tunnel Tour, 1963*

*JJS, Robert Peary, Jr. with Ootah and his grandchildren*

Hunter's Igloo

Hunt Preparation

*Greenland Walrus Hunt*

*ACW Station, Project Dog Sled*

*Arctic USN transport*

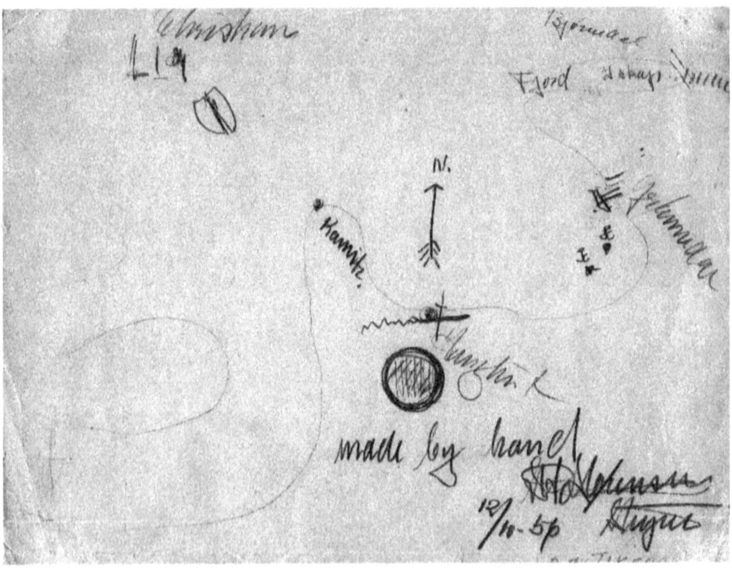


*Arsuk Fjord Capt. Sorenson sketch*

♦ ♦ ♦

# CHAPTER 5

## Antarctica

*Editor's note:* The following is the description from M&E's 1958 report on the Antarctica project.

*This report covers a field reconnaissance and study of the feasibility of constructing an airfield in the Gneiss Point-Marble Point area on the western shore of McMurdo Sound in Antarctica. Investigation and study were undertaken under an engineering services contract.*

*A five-man group from Metcalf & Eddy, together with twenty-one officers and men of the USN CBRU and a photogrammetrist from the Hydrographic Office, conducted the feasibility study. The party was at the site from 13 December 1957 to 14 February 1958 to perform necessary engineering and scientific feasibility studies, including exploration and reconnaissance surveys in the Antarctic area for Deepfreeze III.*

- *Mr. John J. Scheuren, Jr. - Airport Engineer, Group Leader*
- *Dr. Robert Nichols - Glaciologist*
- *Mr. Donald G. Ball - Geologist, Soils Engineer*
- *Mr. John R. Davis - Soils Engineer, Airport Designer*
- *Rev. Daniel Linehan, S.J. – Seismologist*

In early 1957, Mr. Eddy called for me to go to Naval Headquarters to make a pitch for a job the Navy was going to do in Antarctica. The objective was to make a reconnaissance of various areas to develop a location and preliminary design for building an airfield on snow-free land. The firm selected would send a group of engineers and scientists to accompany members of the Seabees to do the job.

The airfield at the Navy's chief base there was at McMurdo Sound. Landings were made on sea ice about fifteen-feet thick sometimes, which was a solid runway for wheeled planes most years. The ice was not constant, and a place to land on solid earth was desired.

We met with the Navy Board–admirals, high ranking civilians, captains, and one young lieutenant commander. The president of the board was Capt. Bill Wesanen, who had been a commander in the Boston area in the Navy, and I had met him casually at the Engineers Club. He was, however, a close friend of many of my close friends at the Club.

We discussed the situation with the Board. I did no talking at first. When the Navy spoke of taking core borings down there, I volunteered that we had rigs and competent permafrost drillers in Greenland and Alaska. When I was told that the Navy was going to do their own core drilling, I withdrew and said no more.

Mr. Eddy thought we had lost the bid. When we got back to the hotel, he began to pack his rig, and we got ready to go back to Boston. He said, "They were cold as ice."

I then told him that, when we were passing through the door on the way out, Capt. Wesanen whispered to me, "How are things at the Engineers Club?"

I told Mr. Eddy he would be needed in Washington to discuss the money and so forth, and that we would get the job. He kept packing and, sure enough, just as he was leaving for the airport, the Navy called up and asked if they could talk contract and money with him the next morning. So, M&E got the job. Later, the young

Lt. Commander Henry Stephens told me I had gotten the job for the firm. I said I had only asked one question. He said it was a good question.

The reconnaissance proved to be an interesting assignment. We had, besides myself as party leader, Don Ball, M&E's top number one soils engineer (I should say *soils specialist*—I don't like to say expert); Dr. Robert Nichols, geologist from Tufts University, who had been in the Arctic and the Antarctic many times; Dr. Daniel Linehan, a Jesuit and a world-class seismologist; and Jack Davis, an excellent airport designer who later became Chief Engineer for the Mass. Aviation Authority. We also had another man with us, sharing our Jamesway hut on the edge of the ice cap. He was a surveyor, a geodetic type from the Navy hydrographic office who was put in control of the aerial photographs that would be taken by the Navy.

There were twenty-one Seabees commanded by Lt. Commander Stephens. There were surveyors, equipment operators, and general construction stiffs. There were also two lieutenants JG and a warrant officer.

We had a couple of D9s, then about 44 tons–about the biggest bulldozers around, compressors, and Gardener Denver track drills, a cherrypicker that was track-mounted, and some sixteen-yard scrapers to be hauled by the D9s, plus two four-wheel drive pickups and riggers and so forth. This equipment was going to be sent down in a ship. We also had pumps, diesel electric generators, even a couple of electrified toilets that would burn the waste, plus stoves.

We all met at Quonset Point, Seabee Headquarters in Rhode Island and flew out in a DC6 Navy aircraft. Some of the enlisted men were having fond farewells at this point. One woman had on an evening gown and was tearfully bidding her man goodbye–he was the cook, Sam Bass, and a good cook, too.

We stopped several places along the way: San Francisco, Hawaii, Canton Island, Fiji, and New Zealand. We stayed about two weeks at Christchurch, New Zealand, while awaiting transport to the Ice (as they called it down there). We stayed at the Clarendon Hotel, the hotel where the Queen stayed. In a park outside stood a statue of Capt. Robert Falkland Scott, who was second to the Pole after Amundsen and who had perished on the way back.

New Zealand is a lovely place at that time of year–Antarctic Spring. You could swim at the beach in the morning and ski on a glacier in the afternoon. Life there consisted of getting up, walking around the Avon River, stopping at a small store to buy some Spanish grapes, trying to find food that was not cooked in mutton fat, and watching the daily swill. At that time, the pubs were only open for limited hours, so from about 5:00 to 7:00 in the afternoon, men would swill all the beer they could hold and then go home when the bars closed. If you lived in the hotel, there was a lounge where you could drink later. There was no tipping in New Zealand, except in these lounges. You could not get coffee at the dinner table. You had to drink it in the lounge.

For some reason, the toilet paper at the hotel was like butcher paper, very hard and not absorbent. One of our people shocked the manager and asked if the Queen had to use that paper when she was there… He was stupefied and made no comment.

One day, I went down to the warehouse where the U.S. Navy had all its supplies and got a carton of toilet paper–good ol' U.S.-style. Because of the size of the box, I took a taxi back. The fare was a florin (2 shillings). I gave the driver a half crown, which was 2.5 shillings. He looked at me and muttered in such a way that I asked the desk clerk when I came back if a half shilling tip for a 2-shilling ride was not enough. The clerk looked shocked. You're not supposed to tip at all!

Finally, we left. By this time, the enlisted men who were staying at a different hotel had all found ladyloves and hated to leave, taking phone numbers from these girls. A Seabee is, of course, a sailor as well as a construction stiff and therefore is doubly sure to be able to handle any situation.

In the British Commonwealth you cannot, by law, patch in telephone calls by radio to local phones. My family was in

Bermuda, and although I could get calls into the airbase there, they could not patch it in. They had to bring the message over from Kinley Field. But the ham radio Father Dan had in Antarctica could patch in calls to the U.S. These sailors, these Seabees, did it all the time, talking to these New Zealand girls.

Incidentally, these New Zealand girls were so pleasant and the atmosphere in the country so delightful, when the icebreaker *Glacier* was leaving New Zealand for Boston, Captain Sam Houston gave the crew a choice: stay two weeks in New Zealand and then sail directly home via the Panama Canal to Boston without any stops, or travel up the east coast of South America and stop in Buenos Aires, Rio, and so forth. The crew unanimously voted to stay in New Zealand for those two weeks.

We flew down to McMurdo on a C124 that was officered by Capt. Rinkle and Capt. Burnett. Then, some of us transferred to Marble Point by helicopter and some by small ski plane on the bay ice. It took these people longer because of the tide crack. We set up the Jamesway huts and had a small city. We dammed one of the melt streams from the edge of the glacier to get fresh water and to set weirs, so we could measure the melt amount and also study the possible use of ice in any future fill for runway embankment.

Before we received the equipment, we took various surveys to locate a runway alignment and dug lots of test holes to locate sources of fill material and estimate the ice content on the soil. We had unloaded from the icebreaker *Glacier* plenty of shape charges, 15-pound and 40-pound, and some half-inch blocks of TNT and boxes of C4, a plastic explosive like TNT. It was safe except for high explosive caps but also a little stronger and easier to handle because of plasticity.

After we blasted a hole, work was required with shovels to trim the side of the hole. This work was not appealing to the Seabees. Father Dan, SJ, was busy with his seismic work, and he had a couple of Seabees who could handle explosives for vibrations for his refraction shots, to check the depth to rock.

The Navy had sent the warrant officer to a school that taught explosive handling. They had taught him that, when he was making primer and crimping a blasting cap for a fuse, it should be done behind the back, so if the thing exploded, it would not harm your face or eyes but hit you in the behind. I did not agree with this procedure, but when we wished to fire a 40-pound charge, the first charge at the site, I told the warrant officer to set the charge in a wooden frame instead of the metal frame that came with the charge. It was okay for demolition pillboxes. We used a wooden frame because the steel would fly through the air like a scythe and clip anyone in range. We also prepared to get behind a small hill, so we would be in defilade when the shot was fired.

The officer put the fuse into the cap behind his back, then he pulled the fuse lighter to light the fuse and fire the charge. We walked away and lay down in a little ravine and nothing happened. I looked at my watch; no one spoke. After waiting twenty minutes, I remarked that it was no doubt a misfire and would have to be fired again. No one moved.

Of course, the warrant officer was responsible. He had to decide that it was a misfire. After a half hour, I mentioned out loud that someone should remove the blasting cap from the well and the shape charge and replace it with a new cap with a proper length of fuse lighter. No one moved.

So, I went up to the 40-pound charge, took the fuse and cap off, and placed it underneath the shape charge. Then, I put together a new cap, fuse, and fuse lighter, taking care the fuse was cut squarely at each end, that proper contact was made between the cap and the fuse, and that also the fuse lighter had the fuse in. This charge went off.

Afterward, Cmdr. Stephens said I should fire all the charges. This suited me fine. Of course, Crosby, Stockton, and Father Dan had their own group, but they only fired light charges for seismic work, and I noted they did not follow the erroneously described plan of making the primer behind their backs.

The only time anyone set off a major charge after that was when the Commander and his officers decided to blow up a bergy bit that was on the shore and could interfere with the ferrying of the equipment from the cargo ship to the beach. This operation entailed compiling a lot of satchel charges around the ice and hanging some on the sides. It blackened the ice but did not fracture it.

No one said anything. I could have destroyed the berg by firing about three 15-pound shape charges with extra standoff and then filling the ensuing holes with C4 and firing again. We did not need to clean the beach any better.

When the ice broke up and the *Greenville Victory* could put its cargo over the side, we brought all the cargo ashore. First, we brought the D9s on barges and unloaded them in three feet or so of salt water. Then, we pushed sand and gravel from the beach out into the water to make a couple of unloading docks. The material and equipment were unloaded onto these jetties.

As I had suspected, the core drills, which were supposed to be cooled by air, were not properly cooled and could get no samples, because there was no core drilling fluid. In the Arctic, we cooled the fluid by either ice water or heavy doses of calcium chloride or used diesel fuel that was cooled by the outside temperature. Fluid was collected in the mud pit and settled out and pumped to another 3,000-gallon tank, which would be alternated with the first.

In this case, because the outside temperature would be -30°F or so, the driller and his equipment worked inside a heated shelter or wanigan. The next year, when M&E had extended the contract, we used ice from the ice cap and water and calcium chloride to keep the fluid cool enough to get frozen cores to determine the ice content of the soil.

However, with our test pits and seismic work, we drilled some holes with our percussion drills. Wagon drills mounted on Caterpillar tracks were powered by an air compressor that the crawler drills towed. These holes, of course, did not supply cores for ice content, but by observing the drill cuttings that were blown out of the hole, we could get some fairly accurate information for the types of soil, the depth to sound rock, which much to my happiness and Father Dan's coincided exactly with these seismic depths.

Naturally, we had the usual amateur who wished to play with the equipment. The aforementioned warrant officer decided to drill a hole with the new equipment. He managed to hang up the steel. I could have gotten it out by pouring some hot water down the hole and racking the steel up and down, but the Commander wished to have his grievance and said he would make the officer pay for the lost bit and steel. The value, at that time, was about $38, but I don't believe he collected it.

In WWII, the Seabees were almost all skilled construction workers and could operate all equipment. These Seabees in 1957-58 were new to the business. However, I taught some of them how to run the crawler drills. They did not need any instruction in bulldozers or using the D9s for hauling scrapers or pulling rooters. After we had our equipment, we were able to open up some areas for potential burrow pits, quarries, and so forth. We also built a small airstrip. The first, we were told, in the Antarctic.

We had some Otters (single engine aircraft) visit us from McMurdo. We were visited by Admiral Dufek, who commanded for the Navy in Antarctica. Because of the Geophysical Year, there were many scientific bases around the Antarctic. Each had a civilian scientist in charge of scientific work and a Navy officer who commanded the Navy people who were in charge of all the logistics. We were not part of the Geophysical Year, and our setup was a little different.

Our work there carried a very high security clearance for a while, until Dufek broke the news to a group of newspaper reporters who were stationed at McMurdo, in quarters known as Bullhime. Of course, Amundsen, when he was there, called his camp Framhime after the ship *Fram*. Dufek brought the conqueror of Everest, Edmund Hillary. Hillary who was preparing to leave McMurdo and travel to the Pole, setting up depots along the way for Vivian "Bunny" Fuchs to use on his way from the Pole. Fuchs was then traveling from the eastern side of Antarctica and would touch in at Amundsen-Scott Pole Station and then cross the ice cap to the McMurdo region, using the depots set down by Hillary.

I found I enjoyed Ed Hillary very much. He was a New Zealander, a New Zealand farmer, and a beekeeper. He stayed close to me while he was at our camp and watched the performance of the D9s, bulldozers, and ripping and scraping, etc. He wanted all the details: horsepower, weight, production. He had tracked his equipment to the Pole, but of course, none were anywhere near as large as the size of these big brutes. Our wagon drills interested him, also. After the visit, someone decided to lift the security, and Dufek returned with several planeloads of reporters.

These reporters had been mostly cooped up at Bullhime and were hard put to get out a story each day. Scientific studies take time, and there is not usually any daily information worth reporting. My wife and family were in Bermuda at the time, and she found out a few things about my presence in Antarctica. Normally, our work was not the subject of stories in the *New York Times.*

Dufek had a meeting in our mess hall. He introduced me as the leader of the expedition and then introduced each member and described each of our duties. Ball, Nichols, Davis, Linehan and I. He made it clear what our mission was. Then, some of the reporters asked questions.

I remember being asked why M&E had a group here. I answered that we had been and still were the Architect-Engineer on the Thule airbase, which of course had been publicized five years before. I described our work in Antarctica, where the runways should be located, proposed water supplies, our soundings for docks, possible locations for hangars, and estimates of possible interference by the ice cap and of the surrounding hills and so forth.

When Becker of the *New York Times* published it the next Sunday, he wrote that Father Dan was engaged in seismic work in the area and Dr. Nichols was fixing the explosives for him, and that there were several engineers with him. My wife found out where we were. In contrast to the *New York Times's* sloppy reporting, Noel Barber of the *London Daily Mail,* who had already written a book about the southern ice–an accurate one–wrote in the *Daily Mail,* the "Dream of Marble Point," in which he described the location of the proposed air field and speculated how this would open new routes between Southern Africa and South America and New Zealand and Australia. It was a splendid performance. We have corresponded a few times since.

There was considerable guessing about there being a race to the Poles between Hillary and Fuchs and that they were competitors, even enemies. The truth, of course, was that they were the best of friends, and Hillary had less ground to cover and no scientific investigation to perform. His job was only to get these depots together so that Fuchs's passage from the Pole to McMurdo would be easier. One can only remember Capt. Scott reaching the Pole and then not finding his depots on the way back and perishing.

We completed our work and had only to bring our information back to Boston to incorporate into a report to our client, the U.S. Navy. Admiral Dufek had other ideas. He wished us to look at a piece of coastline near Cape Hallet that he thought might be feasible for an airfield. Cape Hallet was a IGY (International Geophysical Year) base and was inhabited by U.S. and New Zealand people and penguins.

We had only seen a few penguins at the Marble Point area, but there were hundreds of skua gulls. These birds were thought to make most of their living on penguins. They were a sort of vulture but also take live prey. I suppose Marble Point was a breeding ground for one of them.

We found some with aluminum and red bands on their legs. Dr. Carl Eklund, a great ornithologist who had spent much time in the Arctic and Antarctic, had developed a theory on how to count penguins in a colony. He claimed that you just counted the skuas, multiplied by twenty-five, and there was the number of penguins. I had not heard of this, but a year or so later, I met him at some cold-weather conference in Cambridge, and I asked him what these bands on the birds indicated.

He said those colors would have come from McMurdo. Then, he asked me how many skuas were at Marble Point. I told him 350, plus or minus. He asked how many penguins, and I had to tell him none, only one Emperor and three Adélies were all we ever saw. He walked off and never spoke to me anymore that day.

We later met on the Greenland ice cap; I lent him some tweezers for some bird skinning he was doing on an Arctic tern. He thought, and I think he was right, Arctic tern and Antarctic tern were the same species. Karl has long since passed away, but we parted good friends and never mentioned Marble Point again.

Back to the penguins at Cape Hallett. The Admiral wished Commander Stephens and I would sail on the icebreaker *Glacier* and stop at Hallett to look at it as a possible runway. The rest of the M&E team and the rest of the Seabee unit would fly back to the U.S., while Stephens and I plus one of his surveyors would look at Cape Hallett. There were about thirty-five people at Cape Hallett when we came ashore from the *Glacier*. There were also perhaps 1,500 penguins.

The people both New Zealand and U.S. had when the station had been set up had put up fine chain-link fence to separate themselves from the penguins. However, the winter before, there had been a heavy snowstorm, and the fence had been covered with snow and then it had blown down, so now the Kiwis, the New Zealanders, and the Yanks all lived together with penguin remains and penguin droppings. All this had built up over the years; the stench was almost a physical shock, and the noise of the penguins assailed the ears. Stephens and I had looked at the charts. The only place a runway could have been built was across the litter of penguins, and it was only 1,700 feet long.

Before we went ashore, I proposed to shoot a 40-pound shape charge to check the soil underneath the penguin remains, while Stephens and the enlisted surveyor laid a line and took a profile and also measured the clearance angles at each end, to check interference with any flight plan. The *Glacier* had, of course, plenty of shape charges, the stands for which were not high enough but were also made of steel. We needed a higher stand and of course could not have a piece of steel slicing penguins and people across the area.

I made a sketch of a stand to be made of wood. The Navy had one officer, Lt. JG, who was getting, I think, $50 extra a month for handling explosives. When we came ashore in LCVP (Landing Craft Vehicle Personnel), I found that the stand was beautiful, but it was all made from angle iron, so it could not be used. Fortunately, I found some old orange crates in the dump and stood two of them on end and set the charge on them.

When the JG came up with his blasting cap and fuse and a fuse lighter, evidently, he had not taken the course that taught about priming behind your back. Then, he fixed it so we were all at a safe distance. The resulting hole could not be completed in the same manner as we had been used to. The material around the penguin litter was a relatively fine-grained gravel that was loose and wet and contaminated as far as we could dig with the polluted matter from the penguins.

We look the samples. By this time, the surveyors were finished, and we went back to the ship. When we got alongside, the captain of course was the first up the ladder, then Commander Stephens and I. As Stephens and I walked toward our room, we started disrobing and throwing our clothes over the side. By the time we got to the room, we had no clothes, just a bag of soil samples. Then, we showered and put on new clothes. The samples we did not keep long. They were so saturated and stank so bad, we threw them over the side also.

The trip to Christchurch was uneventful, except we couldn't make very good time, because a few days before we left the ice, the *Glacier*, with the captain asleep in his cabin and a young JG on the conn, hit a floe of rafted ice and split the bow (1-5/8-inch stainless steel plate), and the radio mast fell into the stack. The *Glacier* was brought to Marble Point. We picked up some sand and gravel there, which we mixed with some cement and made concrete to fill in the bow area, after the stainless steel was welded.

There was going to be a group wintering over at Little America, a camp built on the ice shelf that Byrd had had as headquarters. This camp was gradually losing ground, eaten away as the ice shelf moved toward deeper water. I suppose by now it is long since gone into the sea.

The officer to be in charge at Little America that coming winter was Capt. Pat Maher, a Navy captain. He had been an icebreaker captain and was very experienced in the cold and ice. He was aboard the *Glacier*, as well as the other captain, a medical officer, a Dr. Slagle who was also going to winter over. We, Stephens and I, were eating on the flag mess of the *Glacier*. Capt. Houston, of course, ate in his own quarters.

After we got the aggregate for the concrete for the *Glacier's* bow, these winter-over people wanted a sauna, a steam for Little America. Of course, there would be no gravel at Little America, so I picked up some proper rocks on the beaches of Marble Point to make the sauna.

You need cobbles similar to those for a New England clambake, though of course, for a clambake, you only use them once. I never had been sure they could be used again, but no one wants to take the chance. On the *Glacier* we had some fine rocks to use in the sauna. We designed a setup so that, inside the sauna room, they would be set upon a bench backed up with steel sheet inside, and on the outside, a flame would be applied to the steel. The rocks could be dry heated or, if you wanted wet heat, you'd put a pail of water on it.

The sauna room was in an insulated structure under the snow; an iron ladder extended to the top of the snow. These characters would get the fire hot outside, and when the rocks got hot, we would throw water on them and then have a steam bath. Sometimes, they would increase the heat by not using any water.

Maher, Slagle, and the chaplain there used to get into this room and try to make the 300 Club. This was the Club who would get the temperature up to 240° inside and then run up the ladder and roll in the snow at -60°F outside. That was the 300 Club. We left the rest of the people who were wintering over along with the sauna rocks. Slagle and Maher rode with us to New Zealand and were going back on another ship.

There was, of course, no alcohol allowed on board, but just before we left Marble Point, Father Linehan had arranged to have a case of whiskey shipped to Marble Point from a Navy wine mess in New Zealand. I think it cost $1.10 a bottle. We didn't get to drink it all at Marble Point, so Hank Stephens and I had several bottles: five each to take through Customs and another for the passage. The one for the passage, we put in some plastic bottles for water samples.

We also had some bottles filled with heavy brine we had discovered partway down one of our drill holes. Bob Nichols was to do some work with this later at Tufts. One night, as we were preparing for our pre-prandial cocktails, all the lights went out. Instead of bourbon, we had some brine (not for long).

One of the greatest things on that trip was a New Zealand photographer named Derek Wright. He had been with Hillary. He used to claim he was the only man to walk backward to the South Pole, taking pictures of Hillary all the way. He also regaled us with tales of the trip.

I had wanted to get as much of the report written in Antarctica as possible. However, the Navy decided we should leave earlier. Father Dan was the only one to have his report in my hand before we left the ice. It is hard to discipline most scientists, but I suppose the discipline of his religion and his order made him different from the rest. Dan used to tell us how he had to cut hair and also take care of pigs when he was in training. Before we left Antarctica, he gave us all a good haircut.

Just before we left, I found a piece of plywood and wrote the following on it: (Nichols and Linehan said it was like how Shackleton, under stress, wrote poetry; though I was under no more stress at that time):

*Farewell to Marble Point – I part with you as oft before I've parted with good friends and sweet, and now I shake forever more thy memories gold dust from my feet, farewell, farewell.*

Probably should have said, "Memories snowflakes from my feet." This I had paraphrased from a poet named Thompson, who served in the rebel army in the War Between the States. I never knew whether someone else wanted the verse or needed firewood or a piece of lumber, because the next season, Don Ball and crew used the same hut, and the plywood and the poem had disappeared.

As soon as I got back, I headed into the BMEWS work that I had mentioned earlier and got going on the C123 and C130 work on the Greenland icecap. I had two small problems; one of them was Dr. Bob Nichols. He wanted extra time to produce a longer paper on the brines at Marble Point. Professors have to publish or perish all the time. He is one of my dearest friends, but as I say, scientists are another kind of people.

I had made an overlay of some aerial photos to show the Marble Point area as seen from the north. Of course, the custom is always, even at the South Pole, to show the north arrow at the top of any drawing or map. The old draftsman, Scotty Webber, who had been fifty years at M&E, wanted to put an arrow on it, but he wanted to invert the drawing. We finally settled the matter by not using an arrow at all, and Scotty's professional pride was intact.

In Antarctica, we had two navy photographers. One was a warrant officer, Conger; a man of long experience in the Navy, especially in Antarctica. His assistant was an enlisted man.

One day, the Navy commander (Lt. Commander Henry Stephens (Hank) and I climbed the glacier to visit the camp we had on the Piedmont Glacier. Hank and I just carried Musette bags, ropes, and ice axes. The two photographers went with us, one with a movie camera and one to take stills. The enlisted man had the 16mm movie camera.

We had already been up there and flagged the crevasses and so forth. All the way up, the enlisted man was shooting at every step that moved. Conger only took a few still shots. We stayed on the ice cap that night. Conger and I slept in sleeping bags in an Arctic two-man tent, a little more complex than the old "pup tents."

When we got back to the camp, the enlisted man found that he had left the door open on his camera and got nothing. He told Bob Nichols, who was already living at that camp. Bob said that telling us was his second mistake that day, the second thing he'd done wrong.

The next day, we returned, leaving Nichols, Ball, and two enlisted men up there. They had been brought up a week before by helicopter with tents and stoves, even some skis. I decided to ski down on the way back. There wasn't enough slope for a real run.

The enlisted man photographer kept telling me, as he trotted along beside me, that he was quite a skier and went in the ski trains out of Boston to New Hampshire. I let him have the skis and poles, and he shot off as fast as he could go.

He had forgotten that on the way up we had to cross a couple of shear planes with accompanying rocks and gravel and rough snow that occurs in these places. We saw him about a mile ahead of us go flying in the air as he hit the gravel. When we caught up, I didn't offer to carry the skis or poles, so he had to lug them all the way back down the snow of the glacier, all the time looking at me with sad eyes like a puppy, like, "Why you do this to me?"

When we had a few drinks, which was seldom, Conger, who knew his Antarctica history, used to talk about Ponting, who was Scott's photographer, who carried a wooden camera he had made. It was better than steel in the cold. Dick used to say, "If I had a wooden camera like Ponting, I could have been a greater photographer."

Ed Hillary, as I said, visited our camp. For some reason, this photographer, a New Zealander named Derek Wright, came back by ship with us. Hillary had flown back. Wright was a cocky little fellow. He was in the Flag Mess with Hank Stephens and me. He used to regale us with tales of his trip from McMurdo to the South Pole and return.

On the subject of some small tractors that Hillary had, many people thought they were inadequate. Derek used to say to me, "They were the best little crevasse finders you could ever find."

When Dufek landed at the Pole in his ski equipped R4D (Air Force and Army called it C47), we used to joke when picturing this feat of the first man to land at the Pole by air, because Conger was really there first. Dufek jumped out, then Conger jumped out, and then Dufek popped back in and then jumped out very photogenically.

All the photographers worked for the U.S. Army, Air Force, or Navy on our Arctic and Antarctic work, but they were always partial to our Arctic engineers—people to show them places to photograph. They found out that we knew what's new, so that's why we were most friendly with them.

And you always have liars. The Inuit, who never lie, called Peter Freuchen, "Peter the Liar," but there is really no word for liar in our sense. Peter the Storyteller is probably more accurate. He told stories that they might not have believed, but they weren't harmful to anyone. Anyway, Peter told some tall tales, as we say. The story should not lose anything in the telling or not be embellished.

On the Greenland work, we had two excellent liars—Joe Welsh and Arthur Alexander. Lou Litchfield used to say, "Joe, you tell these lies so much, you believe them yourself."

Joe would say, "Some of it's true."

That is like Boswell, who told Johnson, "You have to believe only half of what so-and-so says." And Johnson asked, "How do you know which half to believe?"

Anyhow, Alexander would tell some wild tales, too, and when someone told him off, he would say, "You don't have to believe everything I say."

One of Alexander's tall tales was told when the Corps of Engineers had a briefing; I wasn't present. Alex described the situation at Cape Christian. Major Stone, an Air Force officer, said, "As a matter of fact, Mr. Alexander, you weren't there at all."

Alex shook it off. "Mr. Scheuren was there, and he reported to me. There was no need for me to be there." This was true, but it would have been better if Alex had not been on the plane at all, and I could have had used the place for one of my surveyors.

Then there was Dr. Freistrup, a Danish geologist. He came into our camp at Kulusuk one day from field work on the Angmagsslik glacier with just an ice axe in hand to make a cane, along with a rugged Danish student and assistant (female) carrying sleeping bags and so forth. She told me it was lovely out there, but she was happy to get into our shower room and wash up.

Henri Bader used to remark (I don't believe he ever tried a girl in camp) that when you are living in a tent in the snow doing field work—no one uses the word *exploration*—at first you get used to the smell of your fellows, and then you get used to your own stink, and then it's okay.

We had a job in Thule to check groundwater flows. We decided to use Iodine-131 as a tracer. Iodine-131 is a radioactive isotope with an eight-day half-life. We went to a school given by Tracer Lab to familiarize some of us with the use of the radioactive material and get an AEC license to handle this material. Our party was Dewey Moody, the engineer, his brother a chemist, Fritz, who was the assistant engineer, and Freddie White, who was the handyman. The two geologists from the Corps of Engineers and me took the course. All passed except one of the civil servants. We set up the program, which included a small hot lab for testing, etc. This lab was off limits to all except our group who had AEC approval to handle the materials.

We dug pits and dumped in a small amount of Iodine-131 in each hole and dug pits downhill to sample and record the concentration and rate of flow. The program was very successful and led to the setting up of design modifications to the hangers and taxiways, which were very valuable.

The program had one very minor glitch. Fritz was eighteen; the exposure allowance allowable was thirty, with the formula = 30 plus a constant times age minus eighteen, millirems a week. 30 millirems, therefore, was all that Fritz could have, so he had to be assigned other duties so that he could be exposed again the next week.

There were always hassles, because visiting "firemen," officers, and so forth were anxious to get into the hot lab. I don't know why. Besides the engineers, they sent up a safety engineer who went upstream along this melt river from the glacier with a Geiger counter and came back excited. He thought that dumping the spent material was causing the area to have high radiation. He did not know that Thule is at the point on the Earth where cosmic rays concentrate, and the background radiation there is much greater than in New York. As a matter of fact, the place where we usually dropped the "spent" radiation was way downstream from where this clown had read his counter. We kept this spent fluid in drums for a few weeks, and at eight days' half-life, which it never came in at anyway, it was only a few weeks before the residual was negligible and could be ingested, if one wanted.

The Iodine-131 came in drums labeled with the atomic warning. Although it was packed so no radiation came through, the marking of course terrified many people. Several times, we couldn't find a shipment because the Air Force had placed it all by itself in an empty hangar, and by the time we got to it, it was all run-down. Half-life means that, in eight days, one half is gone and only one half is left; in eight days more, half is gone again. So, pretty soon, you get down to one-sixteenth, which is nothing in that particular isotope.

The work on the DEW Line Extension was continuing. I had taken off to Marble Point right after I turned in my survey notes on Dye 1 and Dye 4, and M&E was loaded with work, so Mr. Eddy had given the job of final design of Dye 1 and Dye 4 to Fay, Spofford & Thorndike. These were people with whom I had been engaged in a bitter battle for seven years with regard to the Fall River Airport. Their unethical conduct on that job had reinforced my low opinion of what that old firm had become.

We were very careful in our Arctic work to be extra selective of any people who had worked for FS&T in North Africa. We did get one or two who turned out okay. Of course, Mr. Eddy was not in the know about this and had FS&T subcontract the design of coastal stations Dye 1 and Dye 4. They had some experience in WWII at BW1 and BW8 and in fact had been our most feared competitor on getting the Antarctic job from the Navy. Those days were gone. In 1946, in Fall River, and in 1957-58, on the DEW Line Extension, they were hardly professional.

The work they did for us on Dye 1 and Dye 4 was not well done. They were supposed to know about the problems of cut and fill in permafrost, but they completely messed up the design, and then we had to redesign these jobs in the field. However, we caught the errors early and the contractors had no complaint, so the work got done at the bid price.

Thus, we proceeded with the building of the DEW Line Extension on the coast of Baffin Island, Cape Dyer, and the two coastal sites in Greenland. We also built four stations in Iceland to cover that link of the communication.

The two ice cap stations proceeded to plan. The contractor had a hard-money, low-bid contract and did okay. We had one problem. The steel that was shipped up was heavily coated with lead paint. Because we were using high-strength bolts, the flaying surfaces of the steel beams and column connections had to be cleaned. We had some extra work burning and grinding off the excess paint. Here again, the Stateside people did not believe us when we said the corrosion would be minimal in these areas and no heavy rust would form. But this was a minor blemish, and the stations are now well over twenty-five years old *(Ed: as of the 1980s)* in service and have been jacked up ninety-seven feet above the original bottom footings and moved laterally several hundred feet.

We were still doing some kind of work at all the bases including Thule, and I will always continue to say Col. Morton Solomon was a giant in all this work. He used to growl sometimes at Mr. Eddy on M&E's prices when we were negotiating new sections of the work.

Mr. Eddy always said, "We give you quality. You can't expect us to give you a Cadillac job for a Ford price."

But one time, the colonel got exasperated about prices for as-builts or completion drawings, and he let his purchasing department find a plumbing subcontractor to do as-builts of the PMT AC&W station for $40 a sheet. They gave him the job. They evidently took the original drawings and copied them and stamped "as-builts" on them.

We were still in Thule when the Air Force let a contract to extend some communications facilities there, and we had to check the contractor. There was an electronics building that, for some electronic reason, had been turned 90 degrees, so the long axis was opposite the original design. The Air Force designers, not knowing that this was the case because it wasn't shown, designed the new utilities 90 degrees off, so that extra cost probably more than made up for what was saved.

As we neared the end of 1961, it signaled the end of our major participation in polar work. When I was leaving Thule, they gave me quite a party—the Corps of Engineers, M&E, and Air Force.

My next project was to negotiate a job with the Air Force, Cambridge Research, for a project called "Waterhole." Even the name of that project was classified. This was to be a terrain-analysis project. The elements were to locate places where aircraft or space vehicles could land in case of emergency: 1—they could walk away for a partial crash; 2—the aircraft could be trucked away by trucks or railroad; 3—the ground could be improved by expedient methods, so the airplane could take off again.

I, of course, had some experience of this in the Army and in the Arctic, and an old buddy of mine in the Air Force had done some great work in the Arctic. Stan Needleman (he and a party) had flown in and landed on the ice at Centrum Lake in Greenland. They had been left there. Then, with wheelbarrows and garden tractors, they had to fix up the ground in such a way that, at the end of the summer, before the ice came again, would allow a C124 to land and take them back to Thule.

We'd have to look at any possible flat place in the U.S. and physically check it. If it was not in the U.S., we could make studies by aerial photos by infrared. I've always wondered how we are always hearing all these tales of the U.S. being stuck for excessive fees and charges. Negotiating this job was a tough one. We had to bid the job on a lump-sum basis then prove that our profit was not excessive and go through everything with engineers and accountants. We got the job and made a total of 7% profit that they had allowed.

The western U.S. has a considerable number of playas (Spanish for beach). This is a geological term used to describe what we might say is a mud flat. Sometimes dry and sometimes covered with water from ephemeral streams. Edwards Air Force Base in California is next to a playa called Rogers Lake. There was another playa near there called Rosemond Dry Lake. Bonneville Salt Flats is a playa. Then there are flat plains and prairies and beaches, like along the coast of Florida.

To gather the information for the study, we set up several teams. Each team consisted of five people: engineers, geologists, and soil mechanic. Each party had two four-wheel-drive vehicles, one an International Scout (like a Jeep) and a Travel-all, which was a four-wheel-drive the size of a station wagon, plus tents, messing equipment, come-alongs, jacks, augers, and hand tools.

The procedures were to check the site, usually by a Cessna 180 conventional type with tail skid that could be landed on a minimal air strip, and then we laid out, with our vehicles, a 15,000-foot, if possible, strip and checked the flight approaches on either end. Then, we steered on a compass course and dropped range poles at intervals. These pits were dug and samples were checked and sent to the laboratory for analysis. We checked the water table with power augers and took photographs.

If there were motels in the area, our people stayed there; if not, we camped out. We had some stoves that used bottled gas. These stoves were the brainchild of young Dave Nassif. He was far ahead of his time and had a whole warehouse full that were selling very slowly.

The security classification presented difficulties because nobody had clearance except Needleman and myself. Some of the men got upset, and one of them was going to write Senator Goldwater about the government money we were wasting.

We had access to small aircraft at Las Vegas, Albuquerque, and Milton, Florida and later on, New Jersey, which enabled us to save time by locating places that were obviously unsuitable without wasting time with the trucks. The AAA was a great help to us. After I had divided up a number of sites for each party, I made an itinerary and asked AAA to make TripTiks, so we got route information. Each party had the best up-to-date information available. We were based more or less in Las Vegas, because we chartered aircraft from Alamo Airlines operated by George Crockett. His assistant was John Matheson, and the chief pilot was Ted Hanson.

Later, after I'd finished the job Waterhole, because they were trying to close out the books on it, they called me up and told me that they had to have me back, because I was the principal investigator, which is very necessary in those jobs. I told Dave I wanted a lot of money, and he paid it, and I flew back to the U.S.

Before that, Needleman and I had worked on the design of a penetrometer, a recording penetrometer, so they wanted to use this. I always said I was one of the most expensive actors in the world. All I did was fly up from Trinidad and fly out to Vegas, rented a car, and drove over the mountains (there was snow on the mesquite), went to one of the jobs, took out the penetrometer, and made the test.

They took a movie of it, and I think maybe I appeared in the movie (for Air Force records) for fifteen seconds. It probably cost them $20,000 for just those fifteen seconds.

♦♦♦

*Antarctica View*

*USN GB-3 Transport Vessel*

*Antarctica Aerial View*

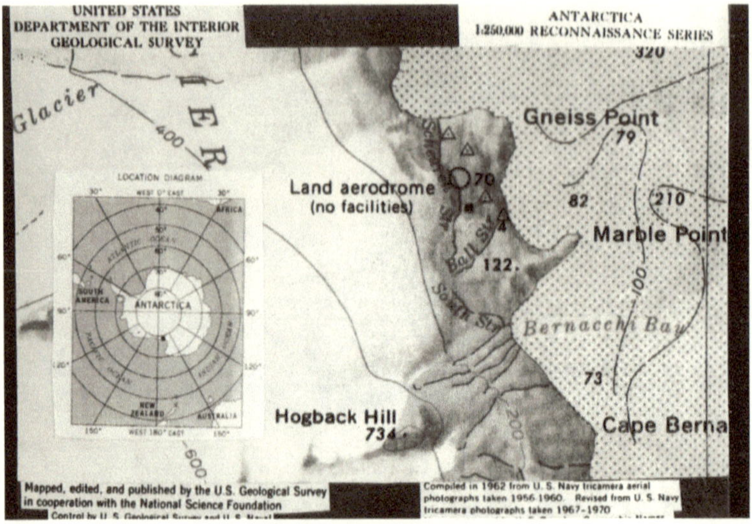


*USGS Scheuren Stream*

*JJS Seabees Drilling*

*USS ATKA Turnpike*

*Hillary-Stephens-Scheuren, 1958*

*Penguin Colony 1*

*Penguin Colony 2*

*Antarctica airfield earth moving*

*JJS with petrified seal*

*Marple Point View*

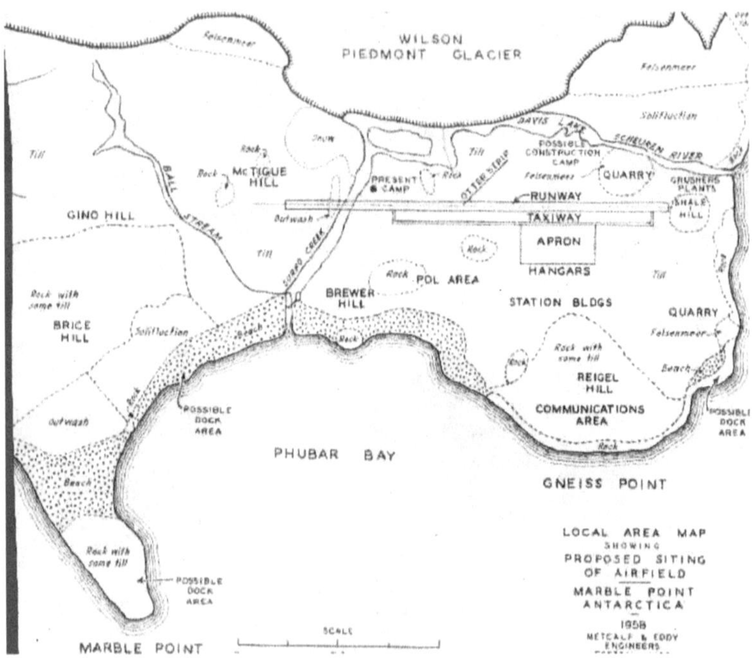


*Marble Point Report Map*

*USN Transport*

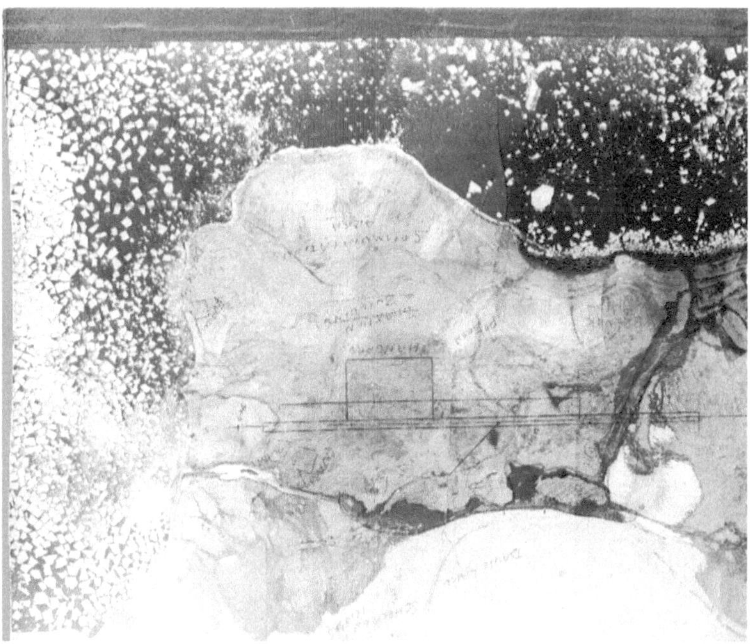

*Davis Lake (top) JJS pencil markup*

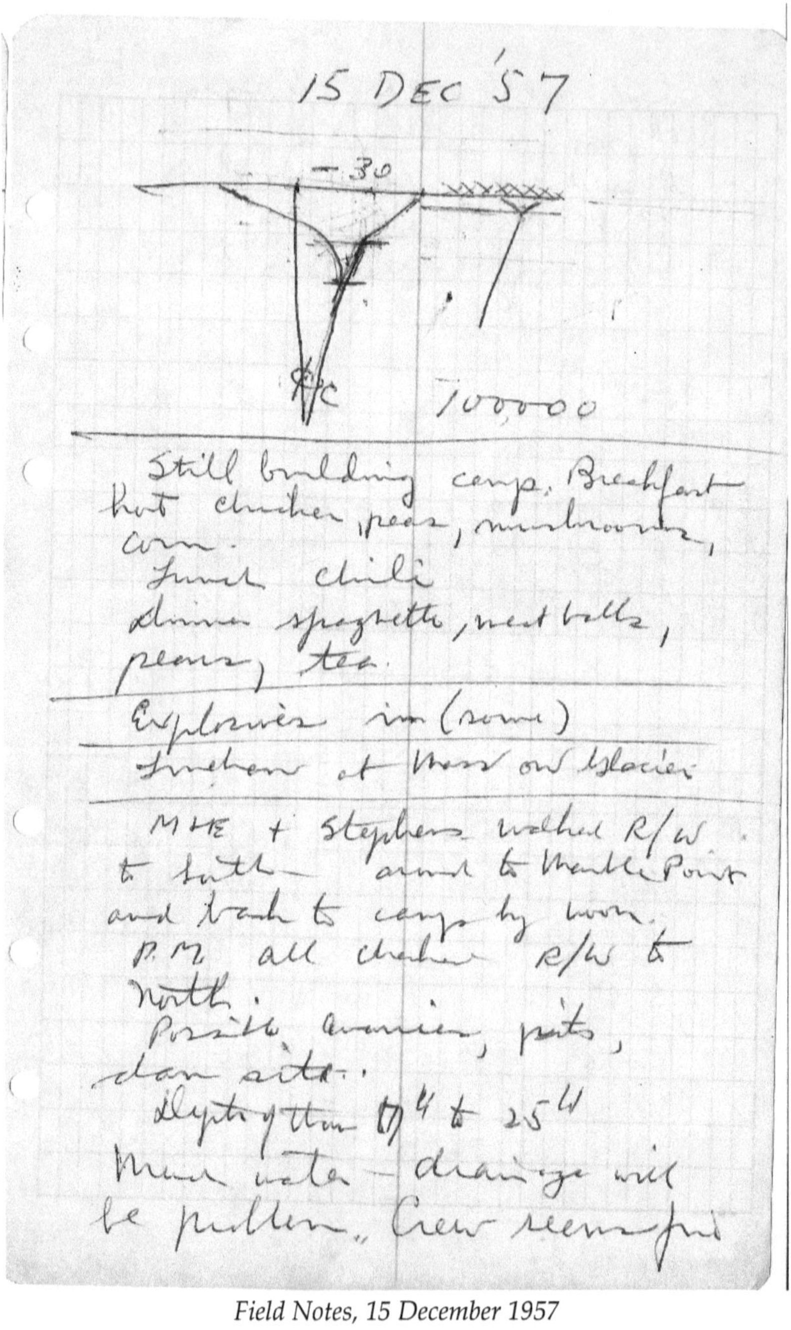
15 DEC '57

-30

100,000

Still building camp. Breakfast hot chicken, peas, mushrooms, corn.
Lunch chili
dinner spaghetti, meat balls, pears, tea.

Explosives in (some)

Luncheon at Moss on Glacier

M&E + Stephens walked R/W to south — around to Marble Point and back to camp by noon.
P.M. all checked R/W to north.
Possible quarries, pits, dam sites.
Depth of thaw 17" to 25"
Much water — drainage will be problem. Crew seems fine

*Field Notes, 15 December 1957*

♦♦♦

# CHAPTER 6

# Trinidad

While we were working in Thule, Greenland, M&E got a job in Trinidad to design and monitor construction of what was then called a nationwide sewerage scheme. I expressed interest in the job. Paton and Eddy believed I was more valuable in the Arctic and sent Sandquist instead.

Sandy had worked in the field and also the office. He had been head of design for the BMEWS project, which had been started while I was in Antarctica. He was a good engineer, but, in my experience, he needed more supervision. He was not always correct in his relations with the client, and in that and other ways needed someone to look over his shoulder. To me it was a question of taking a good lieutenant and making him a captain then sending him out on his own. Montaigne said years ago, "Comrades, you have taken a good captain and made him an ill general." Anyhow, they sent him to Trinidad.

When I left M&E to go with Universal, he had been down there for some time, and everyone was unhappy. His penchant for poor relations with the client stuck with him. The contractors were very pompous and arrogant and cared very little about the people of Trinidad. Sandy and his wife, as well, apparently did not like Blacks, either. In my opinion, when you work in a different society, you can't completely ignore any races of any kind—Blacks, Arabs, French, whatever—because it won't work.

The contractor was arrogant. He would tear up a bunch of streets at a time, so traffic would be halted, and he would not properly backfill or repave, plus would not test the pipes in accordance with the specs. Sandy made an attempt to correct all this, but they ran all over him. And of course, to throw salt on the wounds, he took every Wednesday morning off and played golf with the project manager for the contractor, because he wanted to talk business.

By the late summer of 1963, it had gotten so bad that the client, the government of Trinidad and Tobago, was so upset they wanted to close out the M&E contract, as well as that of the construction contractor, and shut the job down then get someone else to finish it. At this time, I was unhappy with young Dave Nassif's action in holding up payment of my $355 moving bill. When Paton and Mr. Eddy approached me to come back to them, then go to Trinidad and straighten out the job, I was ready. I told old Dave and gave him one month's notice and went to Trinidad in October 1963.

I left my family in Falls Church, Virginia. I planned to come back at Christmas time to bring them down. Paton and I flew down. Shearston, our office manager there who had worked for me in the Arctic, Baffin Island, Thule, BW1, etc., had an apartment in an office building in downtown Port of Spain. He arranged for one for me in the same building.

We gave Sandy his walking papers, and he got ready to leave. Three days later, there was to be a meeting of the Cabinet, the contractor, M&E, Water and Sewage Offices, Borough of San Fernando, and people at the Attorney General's Office to make a last-ditch effort to salvage the job. I was there; Lockjoint, the contractor was there, as was Mr. Bates, who was the technical director of the authority. The Mayor of San Fernando, the Deputy Mayor of San Fernando, a member of the Borough Council, and the Borough Engineer were also there, as well as Sandquist and several members of the Cabinet.

The Attorney General, Mr. Richards, called the meeting to order and stated that it was part of an investigation to see which way we should go. The Minister of Public Utilities, under whose portfolio this scheme was, was Kamal Mohammed, and he asked me directly, "What do you think is wrong? And what do you propose to do to straighten it out?"

I said, "I can see what's wrong. The contractor is trying to run wild, and I intend to put a stop to it."

The project manager for the contractor said, "You've only been here three days. How can you make such a statement?"

And I said, "I've been around the business a long time, enough to know that things are not right. It didn't even take me two days to know that." This pleased everyone but the contractor.

Then, the Minister of Public Works, Said Mohamed, made a sonorous speech and complained that the Borough Engineer of San Fernando, who was protesting the way traffic had been snarled up by the contractor's operations, had called Mike Martin, who was in charge for M&E in that section. They had had a rough exchange, and Mike hung up on the Borough Engineer.

The Minister demanded that there be a complete investigation and that there would be discipline, apologies, etc. The Attorney General said that was a matter between the Borough Engineer and M&E and should be taken up after the meeting.

After the meeting was adjourned, the Technical Director of the Water and Sewage Authority, Bates, the Mayor of San Fernando, Leslie Edmunds, and Deputy Mayor, Sullivan, and a borough counselor, Errol Mahabir, George Davis, the Borough Engineer, Sandquist, and I met— all standing.

I opened the meeting, "Gentlemen, I tell you now, there will be no investigation, there will be no public apology, and no disciplinary action. Every dog is entitled to one bite. Martin has had his one bite. If he does it again, we'll fire his ass."

All was quiet. Davis looked a bit stunned, Sandquist looked fearful, and Edmunds said, "That's right. That's old English law." We all shook hands. The case was closed, and all of them became my best friends.

George Davis used to call me up from New York to pay his respects, and I spent some time with Errol Mahabir in New Orleans at an American Society of Mechanical Engineers (ASME) meeting at which he delivered the keynote address. Errol was a cabinet minister for many years, and he and Davis have always encouraged me to come back to Trinidad and get a great reception.

Bill Shearston, who was always very fond of Sandquist, told me after the meeting that Sandy had told him, "Pitbull John Scheuren, M&E will be off the island in a month." That was October 1963, and we never left until early 1972. It was a long month.

There was plenty of work to do to straighten out the job. Three municipalities were involved: Port of Spain, Arima, and San Fernando. The contractor had spread all around, never finishing anything. I found an easy way to shut down this spread around. I started to enforce the specs, which called for pressure tests on the pipe, and would not allow any more streets to be opened until the pipe had been tested and the streets properly backfilled. I also held up payment to the contractor until this work was done.

This, of course, improved the situation, but the contractor's head office in the U.S. complained to Mr. Eddy and asked for a meeting in Boston. By this time, I had found a proper house and gone to the U.S. to move my family down.

Mr. Eddy and Andy Paton called me up and wanted me to go to the meeting that was scheduled for several days after I had already planned to leave. They asked me what it would take to delay. I said only money. We would all have to stay in a hotel, move my furniture out, and so forth. So, it cost M&E a few dollars. We had the meeting.

I defended my position. The contractor did not do so well. Paton said he could not understand after that meeting why the project manager for the contractor was not fired then, but he wasn't fired then. It took a few months before we got rid of him, transferred out of there.

The Chief Technical Officer for the Authority, Ronald Bates, was a well-educated civil engineer. My wife and I became well acquainted with him and his wife, Anna, who was an optometrist.

Before I brought my family down, Bates invited me to a reception at the University of the West Indies, St. Augustine. It was at the house of the dean of the engineering school. There were some visiting engineers from outside, and as I was standing there (of course, I knew no one there except Ron Bates), a very fine-looking British gentleman came up to and said, "I beg your pardon, are you Dr. Brown?"

I said I was sorry, I was not Dr. Brown.

He said, "I'm Reginald Carr, First Secretary of the British High Commission, and Dr. Brown is a British engineer. I'm supposed to meet him and pay my respects." Then he said, "You are the most distinguished person here, so I thought you might be Dr. Brown." I introduced myself, and we became friends.

As soon as my wife came down, I told her we must have that Mr. Carr to dinner. Carr was an Irishman from Belfast and a great fellow. He got me playing golf; I had not played since 1941. He put me up for membership in the St. Andrew's Golf Club. He said, in the tropics, one must play golf and drink whiskey, otherwise you are in trouble. We kept our friendship together and corresponded for many years.

He was posted to London, and my sister and her husband had lunch with him at his club. Then, he was posted to Atlanta, Georgia, and we visited him there. I was amused when he asked to look at our tickets from Boston to Atlanta—first class—because he was happy to see that our friendship was valued, since I would pay that kind of money to go see him.

I can go no further in talking about our eight-and-a-half-year stay in Trinidad without mentioning my dear Maggie. My dear wife, she has always been more important than I, particularly in our overseas work.

After Eddy, Paton, and I had about completed negotiations on my assignment to Trinidad, Mr. Eddy halted the conversation and said, "Jack, you are going to bring Margaret, aren't you?"

I said of course. She has always been most supportive, I think the word is. When I was charging around the cold and hot deserts of the world, even the so-called glamorous West Indies, she was always taking care of things at home. My advice in 1980 to Peter

Gianakakes of M&E was to check closely the men you send overseas and, as best you could, check into the wives; this is a difficult procedure.

Of course, in Mr. Eddy's case, he had met my wife, not in Greenland, but enough in the U.S. to recognize her strength. Mr. Eddy mentioned bringing Margaret in 1963. When I was retiring from M&E as their representative of the joint venture in Egypt, the representative of our co-venturers there said, half joking, "Jack, why don't you go and leave Maggie here with us?" She had been a tower of strength in Alexandria, helping the young wives adjust and adapt to that culture.

We lived for over six and a half years in Port of Spain, in a large house with walled gardens. As I said previously, Sandy and his wife did not care for the local people, and except for the weekly buffet and dance at the country club where M&E had a table next to the Lockjoint table, there was little exposure to the local scene, either by M&E or the contractor's people.

After I got there, we stopped having that table. There was also the weekly bingo at the Navy Officer's Club at Macqueripe. The U.S. still had a presence there for several years after we arrived; it dated from the old bases for the destroyers' pact of the early 1940s. We found both these weekly assignments very dull and were not as happy as some of our people were with that bit of local culture.

Maggie, of course, registered us as U.S. citizens at the U.S. Embassy and joined the American Women's Club. This got us into a so-called marathon bridge group that the Women's Club sponsored. This was open to all—locals, U.S. people, and other expatriates. Each couple played at their own home with other couples and also at the other people's house. In this way, we met a good cross-section of the local society.

The best thing that happened was we met a British couple, Ken and Poppy Gilmore. They had both been members of the Indian Army and had been married in Burma. He was, in civilian life, an accountant. That's what he was doing in Trinidad. They became our dearest friends.

Reg Carr had already gotten us in the local golf club, St. Andrews. Poppy and Ken got us into the St. Clair Club. This club was one of the last bastions of the British Empire left in the West Indies. 99.9% of its members were British; besides us, there was one other U.S. couple, the Donald Porters, the manager of the local branch of Chase Manhattan, and a Canadian couple, the O'Flynns, who was station manager for Air Canada. The St. Clair Club had a weekly duplicate bridge game that was a friendly game; no one yelling for a director. Meals and drinks were adequate at reasonable cost. No locals were allowed as members.

During the colonial time, which ended only a short time before we got there, the governor, for example, was an ex-official honorary president of the Club. The role of governor had been changed to Governor General, who was of course appointed by the Queen but nominated by the Prime Minister of the country. This system had changed drastically, although the Governor General, Sir Solomon Hochoy at that time, had been a high-ranking civil servant under the Crown. He was locally born of Chinese descent and was not invited to be honorary president or even to be a member. Sir Solo was a very erudite gentleman, and we became great friends. The St. Clair Club finally had to disband and sell out because of the financial situation. Not enough members.

We regretted its closure, but by that time, Maggie had set up plenty of social contacts, and we made out okay. Each year, the club had a St. George's Day dinner, usually held on the tennis courts at the club, and there was always an excellent speaker. Once, they had the head of the British Council; he was unique. He had a PhD from Oxford, and he did not use the title of *Doctor*. I remember he talked a very long time without notes and gave a splendid, not boring, speech—but what he said I could not remember the next day.

The British Council is like our USIS (United States Information Service), and it is barely a branch of the intelligence service in England.

I remember having as guests that evening our friends, Winfield Scott and Rita, among others. Winfield was one of the sons of W.H.

Scott, who had some large businesses in the area, involving building materials, concrete plants, quarries, gravel pits, etc. Old W.H. Scott was in his eighties, and his hands were gnarled with arthritis.

When we were at dinners with him, Maggie used to massage the old man's hands for him. He had been born in Guyana, the son of a bonded servant; immigrants from China who came to that part of the world to furnish the labor that the slaves, Africans, had abandoned when emancipation came in 1835.

At one of his birthday parties, I remember we were the only non-Chinese people there except for one part-Black, part-Chinese man, Aqui Philip, who was what is called a "house child." His mother was a Black servant and his father one of the Chinese sons, not the Scotts, however. This man had risen in the civil service and become Secretary of the Water and Sewage Authority. At the party, after the gifts had been passed out and toasts made, the old man, who was holding hands with Maggie at the time, got up and spoke.

His family name in Chinese had been *Ming*; he started explaining his name. He believed that to make his way in the West Indies and British society, he had to have an English-type name, so the reason he took the name he did was William for William the Conqueror and Scott for Sir Walter Scott.

Then, he went on to tell how he moved from Guyana (British Guyana then) to Trinidad and went into the bakery business, making and selling bread. In those days, the dough and so forth was mixed by hand. Some big man would stir it by brute strength and sweat, which would drop into the mix. So, William Scott invented a power mixer, steam operated by the same furnace that baked the bread. After this, his bread business got really big, and he invested in other essential services, like buying hardware for building construction.

I visited his quarries, and his son, Winfield, who later took over the family business, had the latest down the hole drills, etc. Keep modern was a Scott formula. The old man had three sons: Winfield, named for the old U.S. general Winfield Scott; one named W.H. for the old man; and Miller, who was in the real estate and development business. I always figured the name Miller came from the old family business of making flour.

The Scotts were great friends of ours. Winfield and Rita had two sons, Pierpont and JB Morgan and another named Winfield. Winfield became a senator; senators in Trinidad are appointed by the president.

We met the Scotts at several functions and had entertained one another at dinners at our house or theirs, but it was probably horse racing that really brought us close. Winfield had quite a stable of good horses. In Trinidad and Tobago, there were four racetracks, all turf, and the meetings were short to preserve the turf. There were probably thirty-three racing days a year on the total of the four tracks.

We had never been great racing fans, but one day, I was traveling with Ron Bates, the Chief Technical Officer of WASA (Water and Sewer Authority). We had come back to a visit to a treatment plant we were building at Arima. Arima is Arawak or Carib for water, although it is a very dry place.

As we passed Santa Rosa's racetrack, Bates asked me if I would look in there with him for a while on another subject. It seemed that the track got so wet, the jockeys were refusing to ride, and the owners did not wish to enter their horses. Some of the racetrack people had asked Bates to look into the matter for them. We went in and talked to the Chief Steward and the track maintenance people, and I told Ray D'Abadie, the Chief Steward, that I would look into it. The chief engineer of WASA, Ian De Verteuil, was just about to leave WASA and become a private consultant and, incidentally, joined us on some of our work. Ian and I got together and did some surveying.

I did some soils analysis and drew up a plan to use underdrains, cutoff trench, and so forth. We got it done very cheaply (the track people had very little money; they had been losing it). The plan we had drawn up worked, and the tract became properly drained, even though Winfield growled that, once it had, it became too hard.

I became a sort of hero, and the track published in the local paper that, because of me, VP of M&E, and Ian, the track was now okay. The governors of the track, the head one being Sir Hugh Wooding (who was also Chief Justice of the Supreme Court), gave me and my wife lifetime passes, parking space, and seats in the Governor's box. Maggie and I used to go to the races, mostly at Arima, but sometimes the other tracks.

We used to get good tips from Winfield and won a few dollars here and there. Winfield was always a little discouraged that my wife would be doing some knitting or crocheting and reading a book when the races were going on, because she was completely uninterested. During the time we spent in T&T, we ended up ahead five or six thousand dollars, mostly because, one day, we had a $2,300 win at Santa Rosa, Arima.

The governors were really happy to present me with a check for the $2,300 and said over drinks in the private bar, "See, Jack, we can now pay you for draining the track. We couldn't pay before."

Of course, they got their fifteen percent—the track doesn't lose any money as long as people bet. But this was of double benefit, for the next week I went on leave to Canada with first class tickets provided to Maggie and me by the Air Canada people in Trinidad. We had free tickets and $2,300 to go to Toronto.

Pat O'Flynn, the manager, personally loaded us in first class of a DC8. It was in the rear to the airplane, in propeller days. We rented a car in Toronto and went to London, Ontario to visit our friends, the McKillops. He had been acting director of WASA for a while, and Vern was close to death from cancer. It was good to see them. He retired from London as chief of public works, and they named a park for him.

We drove back to Toronto and flew to Winnipeg. I wanted to look into some large stabilization ponds they were building and also to check into some possible new work for M&E on stream cleanup.

We later got the stream cleanup work, and Doc Sawyer went up and reported on it. Sewage spread consisted of large ponds for oxidation and other large ponds for storage during the winter, when the bacterial action would be nil, so they didn't discharge, they just stored in the winter.

We also looked up the children of our friend, Marsang who went to school there. Carlton Marsang was a Chinese from Canton who had a great restaurant in San Fernando, Trinidad. The restaurant was rickety and cluttered with cases of Coca-Cola and beer and so forth, but the kitchen was spotless. Marsang supervised all the cooking, and his wok was always hot and shiny. Just inside the door were the garbage cans. Each waiter would dump all the excess food into the cans. Nothing was ever reused except to feed some of his animals (pigs and manicou), which he used in some of his dishes.

His daughter, Yikken, was studying at the University of Winnipeg, and his son, Ming, was studying at a trade school there. We invited them to dinner at the old Fort Garry Hotel, which rather embarrassed them, as they had eaten dinner earlier and could not eat much with us.

Carlton Marsang thought this was wonderful—we had taken his daughter and son out to eat in a "beeg hotel." For the rest of our time in Trinidad, he would try to give me food and drink for nothing. However, I would give the money to his wife, who presided over the cash register and knew that, without her, Marsang would give the place away.

We flew to Calgary and drove to Banff and Lake Louise then took the train over the mountains to Vancouver and stayed at the Bayshore Inn there, a great place at the time. Then, we flew to Montreal, where our daughter, Claire, and her husband were living. She had been married in Trinidad. He was studying at McGill, and she was doing medical photography at a hospital.

Maggie stayed in Montreal for a day, while I flew to Boston to check with Paton and Eddy. I spent an hour or so with Andy and heard about three of our oldest friends, one who had just been killed the night before in an auto accident and the other two who were dying of cancer.

After I left Andy's office, very depressed, I ran into Johnny Dennis in the corridor. He was then one of our junior engineers and had been in Greenland. Then he became a partner in a large engineering firm in Florida.

He said to me, "Did you hear about John Raymond?"

John Raymond was senior vice president and in his late seventies. I thought, here is another one, but Johnny laughed and said, "He got married." John's wife had been gone for many years and now, he had gotten married again.

"Where is he?" I asked.

"It's a secret..."

Peg Cocker, our lady in charge of travel arrangements, was sworn to secrecy, but I would not take no for an answer. John was an old friend of mine, so she told me where he and Mary were honeymooning in Bermuda. So, we set off to Bermuda.

At that time, this was a good way to commute to Trinidad, and it kept you out of the madhouse at Kennedy Airport both ways. You could go through U.S. customs there and land in Boston. We got a room at a guesthouse right next to the house we had rented when my family stayed there while I was in Antarctica. Our old house had been made into apartments, and the next house was now a guesthouse. We surprised John and Mary, and they were delighted.

John had only returned a few months from Trinidad, and a month or so after our meeting in Bermuda, he needed some consulting services in Port of Spain, so he and Mary came down to stay with us for a while.

He was seventy-eight then and Mary was seventy-one; they were like young newlyweds. John's first visit to Trinidad was for political reasons. The contractors were in cahoots with the Minister of Petroleum and Mines and had obviously paid the minister off for getting them the job. M&E would never pay any graft, and the minister, Johnny O'Halloran, had asked Mr. Eddy, and Mr. Eddy had put him off.

The contractors had complained to the minister, and he had passed it on to Mr. Eddy that I was not treating them right. Mr. Eddy said he would send John down to look into the matter. John came down and could find no unfairness in the way the contractor was being treated—we were paying him for what he did.

While John was there, I met O'Halloran on the beach one day, and he asked me how we were getting along in the investigation. I told him it was going along okay and that I was sure Mr. Raymond would report to Mr. Eddy. On the way to the airport for John's departure, I asked him what he was going to report.

"I have nothing you don't already know. You write the report and send it to Boston, and I will sign it." I did and that particular gambit failed, but O'Halloran was pacified.

In the meantime, over the years, we got more work on sewage disposal and water supply in Trinidad and Tobago and other locations. Concurrent with this, our position with WASA and the cabinet strengthened. Errol Mahabir had succeeded my friend Kamal Mohammed as Minister of Public Utilities; both ministers were and still are good friends of mine.

At this time, the Electric Commission was increasing its power capacity and had asked for international tenders for an 80-megawatt increase in generating capacity for the Port of Spain plant. This was to be a package deal. The bidders would propose to design, furnish, and install the 80-megawatt power. People from all over the world tendered. General Electric International, which was a U.S. and German company, proposed, as did a company from UK and one from Japan. The British firm was $8 million, GE was $12 million, and the contract was awarded to GE for $12 million.

The Prime Minister asked the dean of the University of the West Indies, Dr. Ken Julian, to look into this matter. His report was supposed to justify this contract. Errol was shaking. Under closest security, he brought me a copy of Julian's report. It was a lot of garbage.

One of the questions, he said he did not have time to go into, and then he did some sort of dance with the figures and heat rate and so forth to cover the angles. I made a critique for Errol, which he still had when I saw him at Houston in 1980. But Julian was a blue-eyed boy for the Prime Minister, and nothing ever came of that thievery.

It appeared that GE got $1.3 million and someone in government got $2.7 million. O'Halloran told Errol it was the Prime Minister, but who knows? (The British firm was Parsons, Ltd.; Parsons, the founder, had invented and developed the steam turbine.)

The last big job we did in Trinidad and Tobago was a report of water for the International Development Bank. This was a $1 million job for us, and it included feasibility studies, preliminary design, estimates, writing up the terms of reference for final design of several dams, pipelines, treatment plants, and pumping stations. By the time we had finished the job, Julian had organized a local group called Trintoplan, in which he and a small body of local engineers, mostly highway and bridge people, would own thirty-five percent and the other local engineers, if we wished to go, would take up fifteen percent of the shares, while the government would take twenty-five percent and the labor unions, twenty-five percent.

We did not wish to go along with it, and it got pretty nasty for a while. Before we had signed the $1 million water study, O'Halloran called me and asked me for Mr. Eddy's address. He wrote Eddy and asked him to come down and visit him. Eddy called me up and asked my opinion. I told him he should come down.

Errol did not ask us for anything for getting us this job. O'Halloran told Errol, "I've always known you were an ass. You are a bigger ass that I thought."

Errol was offered $50,000 by one of our competitors. The man put $5,000 in hundred-dollar bills on his desk as a down payment. Errol told me he had never seen that much money before. He refused it, and we got the job.

When Eddy was coming down, I began to worry whether I had been wrong for all those years about M&E being on the level. I saw lots of books and I'd been on the board, even an officer, and I could see no way any payoff could have been made. But accounting systems get more confusing as years go by.

Then, O'Halloran told me, when I told him there would be no payoff, "That old man knows the score." When Mr. Eddy came down, he went to talk to the minister.

He told Mr. Eddy that I was okay, but he wanted to see Mr. Eddy alone. He asked Eddy for a contribution to the party. Mr. Eddy said it was a violation of the charter and of public law concerning corporate political contributions. He went through the story of Leonard Metcalf and his father, H.P. Eddy, Sr. Then, Johnny began to laugh. He had heard that song before. He had the wrong old man. He said he had to laugh, but we were not to get anymore work there.

The president of American Pipe and Construction Company was named Johnny Edwards. Edwards was part of the partnership that had the construction down there, and Edwards at that time was being indicted in California for price fixing and so forth. He was a noted gambler on horse races and cock fighting.

We did bid on the final design and monitoring of the construction, and as George Davis told me, who was by that time General Manager of WASA, we were low bidder and had the best people, the best proposal, and the lowest price, but the Prime Minister wanted to give the job to local people. The local people were Trintoplan. Trintoplan did not, of course, do five percent of the work; they sublet it to CH2M Hill of Portland, Oregon, who later were and are M&E's co-adventurers in Alexandria, Egypt.

Graft, I'm afraid, is everywhere. Errol was absolutely straight though; I've heard lately he has slipped a bit. Johnny is now a refugee in Panama because of Eric Williams's last appointment. He made Johnny Minister of Horse Racing. A contract was handed to a U.S. firm to build a new racetrack. The firm paid Johnny a lot of money and received a lot of money, but the track never got built.

The government is still trying to get the U.S. firm to return and pay up, but of course they won't. This bribe was over $2 million.

A Chinese doctor, Dr. Ouwai, got mixed up with Johnny and a few other deals, buying some worthless land and putting it up for security and buying some other property and cheating the government. Johnny got the Chinese doctor made honorary consul for Panama in Trinidad; thus, he was able to get Johnny residence in Panama.

When the racetrack thing broke, Johnny fled to Panama. All he left in Trinidad was his home, which, while valuable, did not cover a tenth of what he had stolen. Ouwai also got out on the basis of diplomatic immunity and left his wife at home in Trinidad. The government would like to get both Johnny and Ouwai back, to try them, but there is no extradition from Panama, so they both stayed there.

Last I heard, Ouwai was hurting, and the government of Trinidad would not give his wife an exit visa. She is a Canadian. Ouwai can only practice medicine in Trinidad and Canada. If he goes to Canada, he will be extradited to Trinidad. He didn't get away with anywhere near the amount of money Johnny did, and so he is hurting, and his wife cannot leave.

In the interim, Eric Williams died and the new Prime Minister, George Chambers, was friendly with Johnny, too. After that, Johnny died of cancer. They brought his body back from Panama and gave it a state funeral in Trinidad.

When we were in Trinidad, we were very friendly with Sir Hugh Wooding, who was Chief Justice of the Supreme Court there. He was also head of the Santa Rosa racetrack. He told me one day at races that, if all consultants and contractors would not pay any graft, like M&E, of course there would be no graft. My friend, Issa was in a couple of real estate deals with Hugh, always on the level. Not like Gairy's operation in Grenada.

I did some engineering for them on a hotel project, just speculation—no pay, which was regular on such deals. Anyway, Sir Hugh retired and was made president of the National

Telephone Company and within three months was in trouble because of "bobol" (Trinidad word for graft). His first act was to sell all the company-owned vehicles and rent cars from a company in which he had a large interest. This rather dimmed the radiance of his image.

I remember setting up a means of building small bridges around the country by using flat cars made available by the closing of the railroad. We needed a new bridge for access to the newly-built Arima sewage plant. We bought a flatcar for WASA from my friend, George, who was presiding over the dissolution of the railroad.

I bought it for $100 and, with a few materials and little labor, made a suitable bridge on this access road. Desmond Smith told me this was the only one built in Trinidad, because there was no bobol (graft) possible. When other bridges were needed, designers were hired, materials acquired, and contracts let—no one would ever use those flatcars.

Later, we found that, in Panama and perhaps other places, this gambit had been successfully used, and much money and time had been saved. Compared to the big bobol on Lockjoint, of the 80-megawatt power plant and the racetrack, this was only peanuts.

Bobol, squeeze, dash, graft, corruption—the words are different all over the world, but the results are the same. Graft and corruption equal bobol in Trinidad, "squeeze" in some places, "dash" in Nigeria, but in Thailand, there is not any real word for it. The Thais use the word *corruption,* the only English word among lots of Thai words. We still have, I suppose, some snow removal contractors being paid for snow forty-feet deep, curb to curb, in some cities where, of course, the snow is nowhere near that. Even recently, we heard of Admiral Rickover, that mystical white knight in the nuclear navy, accepting jewelry for his wife from large Navy contractors.

Our work in Trinidad and Tobago started out limited to the sewering of large areas of Port of Spain (POS), Arima, and San

Fernando. Besides the sewers, there were trickling filter plants in San Fernando and Arima and stabilization and oxidation pumps at POS. A separate pumping station was set up at Diego Martin, and a large pumping station at POS to pump the sewage through 48″ force mains to a set of ponds, which were set up at Caroni Swamp. Four of these were twenty-six acres and were called primary ponds, where the sludge would settle out, treated by anaerobic process, and overflow would go into the 220-acre aeration ponds and, after several days, would be discharged into the Caroni River.

We designed and supervised construction of a sewage treatment plant at Mansica College, set up drainage for a race track, organized a system for taking the rum wastes from the Angostura Bitters factory into the system, and designed and supervised grading, drainage, and sewerage for Strickers Village housing project, where we approved and supervised the connections to attach all properties to the sewer system. We worked on another housing development and later made a study of sewage in all areas of Trinidad and Tobago, reported on this, and made recommendations.

We also made recommendations concerning the government-owned plant that made clay blocks, about moving it to a different place and using a better clay, because the clay at that place was not very good, and the clay at Strickers Village was just about right. We also wrote up the study and submitted the preliminary design for a nationwide water system for the whole country of Trinidad and Tobago, writing terms of reference for the final design and construction of dams, treatment plants, pumping stations, and trunk lines for it.

During this time, we worked on AID projects for wells in Macqueripe (Macqueripe Bay), in St. Lucia for irrigation, in Antigua for water supply, and in Guyana for highways. These projects entailed leak-detection surveys and float studies for outfalls. We made a study in Grenada for an airfield there and submitted drawings and preliminary estimates for this in 1964. The

Caracas, Venezuela water people, INOS, asked our advice on groundwater problems there, and we made a visit.

In St. Lucia, we had a chance to see some things that happened to USAID money. The project had been established there between the government of St. Lucia and the USAID for irrigation of 400 acres of banana land. The location of the tract was established by coordinates, down to decimals of a second, latitude and longitude. Some work had already been done in the form of some contour trenches that had been dug at that point, ready to receive the supply.

The Camalles River ran by the side, albeit several hundred feet lower at this point and over a thousand feet away. One means of supply was obviously by pumping the water up to the banana fields. The other was to build a canal from a point about six miles upstream, where the river elevation was high enough to allow the canal to be operated by gravity. The quantities of water had to be enough to produce four inches of water per month over 400 acres, to irrigate the bananas.

We ran a reconnaissance survey up the river and laid out a canal route. It was mostly cut, but there were several places where flumes would be required across some ravines. We also considered use of pipe, inverted siphons, to make these crossings. The other option was to put a pump in the river below the site and raise the water up to the bananas.

We figured the size of the pumping operations and compared it with the construction and maintenance of the canal. The figures showed that the cost of the pump was less at the start, but after about thirty years, the canal would become less expensive.

However, there were some problems with riparian rights that might give trouble from some of the abutters, so we recommended the pump solution. USAID and the St. Lucians authorized the pumping equipment. It was interesting anyway. In the design of the canal, we had figured the size of the canal would be safe from erosion if they were not lined and if the velocity of the water did not exceed two-and-a-half feet per second. This velocity would not erode the banks and existing material and would produce enough water to do the job.

St. Lucia was subject then to a debilitating, sometimes fatal disease called bilharzia or schistosomiasis. The carrier of this disease was a fluke that was a parasite in the snails. These parasites did not hurt the snails, but they got into the water. People who bathed, washed, or drank the water got these flukes into their bloodstream, and the flukes travelled to the liver. The disease is very common in many countries; Egypt's Nile Delta is another place where it is endemic. We knew the disease was endemic in St. Lucia. The son of one of our friends, the manager in Trinidad for Barclays Bank, who came from St. Lucia, died of the disease he'd picked up on a visit to his grandparents.

We knew the Rockefeller Foundation was amassing a great effort to eliminate this disabling disease. Afterward, I read in *Science* that Dr. Ippan of MIT had studied the velocity of water that would keep the snails from breeding. He had fitted little dynamometers on the snails and measured the strength of snail power, like horse power. He found that two-and-a-half feet per second was a velocity at which the snails could not hold their footing in which to breed.

The velocity we, by serendipity, had figured would keep the banks form eroding would be two-and-a-half feet per second in the main canal, but in the smaller irrigation ditches, it would be less. But when we had what we thought was all decided and had ordered the pump, we changed our thinking because of the bilharzia, and decided to use spray irrigation, in which the snails would get smashed up in the nozzles.

We were taken aback by a change in the government of the island. It was then a British colony. The new chief minister called a meeting and accused us of deliberately choosing the wrong spot for the irrigation. I showed him the contract, which included the exact coordinates, and told him they had already dug some ditches. But he was adamant, and the project was moved to another location. The reason was that the chief minister or some of his friends had some property for coconut growing that did not require much irrigation but did need an access road. In order to move the pump and so forth, we had to design a new haul road, so they got a road for their coconuts. But as far as the bananas went, irrigation only lasted as long as the diesel fuel we left in the tanks at the pump house. People would not buy any more fuel, so I don't know what became of the project.

At the same time, the USAID had financed an abattoir for the cattle industry there. Up to then, they had butchered the cattle in the fields. This abattoir had killing rooms, chillers, and storage capacity for frozen beef. There were some problems with the chilling and freezing aspects of this work, which our mechanical engineers fixed up.

Several years later, I talked to some people in Trinidad. One was Alfredo Bermudez, who ran a large biscuit factory there and was in the food business in other ways. He told me the abattoir had never been used for the designed purpose. No one had ever killed a cow there, and the plant was turned into a depot in the Caribbean for "lost meat."

I found that lost meat was meat that had arrived at a port in the U.S. and the trailer was stuck in a corner of a transit warehouse, hooked up to power to keep it cool. Then, when it was supposed to be loaded onto the ship, no one could find it, so the ship sailed without it. The shipper would make a claim and the insurance company would pay him off.

In the next step, someone would find the shipment, buy it cheaply from the insurance company, and ship it to St. Lucia for warehousing and selling. Some of my friends in Trinidad got some of it. I understand it was a big business and that regular traffic was carried on. People could depend on a fixed amount each year by negotiating with the crooks who ran the racket.

St. Lucia is a lovely island. Of course, as Sir Hugh Wooding used to say, it suffers from the unspoken curse of the West Indies, which is sand flies. These savage, biting machines breed in the grass at the edge of the sand and at just about twilight are a terrible curse.

Spraying, fogging, and plowing up the turf along the sand used to help.

The next USAID job we did was in Antigua, which is not so pretty a place as St. Lucia. It, of course, has great weather, but Antigua had a shortage of fresh water. The island is divided into approximately three zones: one is coral; the next, limestone; and the last is sort of volcanic and clayey. There is not much relief on the island, few hills, so there is a great lack of rainfall.

The USAID contribution had been a contract for four wells into the coral. The contractor, allegedly an Irishman, had drilled the wells and left the island. His drill logs were not available, and we were asked to put the wells in operation. Because of the lack of records, we had to make guesses as to the length of drop line required, drawdown, etc. The water, as expected, was very salty, practically undrinkable.

We also looked into the remnants of a dam that had been started at the other end of the island. This area had fair possibilities, and there was good clay for a core wall. However, if a dam were built, the necessary cutoff wall could cut off the aquifer that fed the Bules Brook wells, which were producing good water in that section of the island.

One of the Antiguan engineers had heard from someone that a dam could be built without a cutoff and a clay blanket under the reservoir, to keep the water sealed in. This possibility was out, in my thinking; I doubted the stability of such a structure. I did not recommend it, because an oil refinery was going to be built there that would supply relatively cheap fuel, and a new power plant was proposed. So maybe they could use the waste heat from the power plant to run a desalination plant for drinking water, and also, I speculated, the refinery would need to have some fresh water to operate, anyway.

The refinery was built; the power plant was built, albeit with secondhand equipment. Some of my friends in Trinidad who worked at the refinery would bring home a few gallons of fresh water each day from the refinery's desalinization plant, in order to have some water to drink. Nothing else was done at that time, and then, some Englishman got into building dams all over the country, seven to eight feet deep.

This looked good for a while, because a hurricane brought excess water, but when that splurge was over, they went back into the usual thing of seven feet a year of evaporation, when dams are not much use. The refinery was shut down, but I hear now they are going to rebuild it. As I remember from the figures, use of a hotel guest might be 100 gallons per day and the servants use was 100 gallons per day, so the price of the room would only be a couple dollars a day to cover the water cost.

We had a call from the water authority in Caracas, Venezuela to our Boston office, asking us to send someone to consult about groundwater. I was in Port of Spain, which is nearby, so I went to the Venezuelan Embassy there, got a visa, and flew over to Caracas, 200 miles away.

I met with the INOS people (Institute Nationale De Obras Sanitarier), the Institute of Sanitary Works; they had the Presidente, a large group of water engineers and hydrologists, and one American, an old engineer named Arthur Murrow.

They were a bit perturbed because I was not much good in Spanish. I explained that although Trinidad has a Spanish name, it was an English-speaking country. M&E had done lots of work in Spain and had plenty of Spanish-speaking engineers, but I was nearby, so I came to the meeting. But if we had any future work in Venezuela, I was sure we (M&E) would send Spanish speakers.

They had a sort of two-block situation; there were two industrial cities, Valencia and Maraci, in the west of Caracas. The level of the wells in that area was dropping, and they were hoping to get advice on where to drill new wells near Valencia Lake. It was a sort of closed system; the level of the lake was dropping six feet per year to two, and a prison on an island there was furnished with drinking water by small boats, because of the sulfates in the water.

I suggested that all indications were that the water table was dipping in the whole area, and they should not waste their money having us send some hydrogeologists to the area. (They had more excellent ones than we did.) They should think about tapping a branch of the Orinoco River for bringing in more water. I said my main office would write them on this.

When I tried to go home, it became complicated. I had a reservation on Pan Am that night at 10:30, but when I checked out of the Tamanaco and went to the airport, the Venezuelan immigration people told me I had a commercial visa, and I had to have a tax clearance in order to leave. I hadn't of course earned any money, but it didn't make any difference.

So, I went back to the Tamanaco. All the rooms were full, but I did get a cabana near the pool for the night. The next day, I called the president of INOS and told him my troubles. He switched me over to his secretary, who had what I thought was an improbable name for a Hispanic, Gladys. She got on the phone and talked to another lady, who directed me to the tax office.

At this time, there were soldiers with submachine guns all over the place. I went to the office and met the lady whom Gladys had called. She took my passport and left me alone in a closed room with an armed soldier at the door. After a while, what seemed like hours, she came back with a tax clearance good for six months.

I fared forth again the next evening, as there was only one plane a day. At the airport, I ran into the same grouchy officer as before. He told me I needed stamps on my passport for clearance.

"How do I get them?" I asked.

He shrugged and kept my passport. I went to the Pan Am desk and was told these stamps were only available at the incoming passenger area. I should have obtained them upon entering. They said to go down there where all the passengers were coming out, go up the line, and I'd find the stamp office.

Without my passport and only a briefcase, I started to hit the line. A soldier with a rifle at port stopped me. Fortunately, a sergeant came over, and I put on my best U.S.-officer bearing, polite but firm. Finally, he understood and showed me the way. So, I went in and bought the stamps.

It cost me extra because I had already exchanged all my Venezuelan money and used U.S. dollars; they cheated me out of some change. Then, when I started back, I was stopped by another soldier with a submachine gun. I did not have a chalk marking on my briefcase showing I had been through Customs.

Somehow, I waved and smiled and got the attention of the sergeant, who was about forty or fifty feet away. I held out the briefcase and slapped it a few times. He nodded, smiled, and sent a non-uniformed man to lead me through the mob. I took the stamps back to the officer, who put them in the passport and gave the passport back to me, and I was off to Trinidad.

I had talked sewerage with the people of INOS. The river there in Caracas was dry in the dry season, and the dry weather flow was just sewage. We talked about pumping the sewage to another water course and, by a system of stabilization ponds and aeration, treating the sewage. I also noted a new type of treatment while I was there.

Caracas has a huge shanty town; I remember one of the shanties had large Doric pillars. In the dry season, when the sewage dries up, it used to flow over the whole countryside, including the city of Caracas, and then the settled solids became dust; effective but not hygienic.

There continued to be much opposition to the sewage scheme in Trinidad. It started with Lockjoint's arrogant attitude but was fueled by the need to abandon the soakaways and the need to spend money to make the connections.

I remember the chief of the fire department asking me, "Can you show me where any reduction in disease has been affected by the sewage scheme?"

To which I had to reply, "Can you show any reduction in the number of fires since you've had a fire department?"

The wife of the Governor General, Sir Solomon Hochoy, Lady Hochoy was very concerned about developmentally disabled children. I told her that Cardinal Cushing of Boston had called the homes in his area "homes for exceptional children." They are now called the Lady Hochoy Homes.

I helped there by contributions and by trying to keep her home from flooding. A large drain ran through the house property, and the intake from the stream had been covered with a heavy steel grating and bar screens. Upstream was a slum area; the people living there would steal the gratings to sell. When there was a heavy rain, big stumps and logs would get into the drain and plug it up; the water would flood the home and, of course, cause trouble to the poor children who lived there.

I showed the drains to people one day, when I was soaked to the skin in helping handle stumps and debris out of the manholes. They said, if they could put up new grating, it would be stolen the next day. I showed them how to concrete the bars inside, using railroad rails, and to build a manhole just inside, so the rack could be cleaned from the inside as well as the outside, and the concrete encasement would be such that no one would be able to steal them. Like many other battles I had there, I lost this one, and I suppose it is still the same way.

I participated in several symposiums at the University of the West Indies when I was there. The first was on the further education of the graduate engineer. The British influence in such places as the Caribbean seemed to preclude any physical participation by the graduate engineer. Once degreed, he put on a white shirt and would touch nothing.

My opinion at this symposium was, I'm afraid, developed by a fixed opinion of the British engineers, who only seemed interested in getting good local engineers for low pay.

Another one I participated in was the "Engineer as an Entrepreneur." The dean of the University, Dr. Ken Julian, had written the rules. Papers could be as long as you wanted, but you could not speak more than twenty minutes.

I had practiced by reading mine at seventeen minutes. My mother was visiting then. However, on the morning of the session, the man just ahead of me, an American professor, was not prepared and only had a pile of sheets in front of him, which he picked up one by one and ran over his time so much, I only had a few minutes for my own reading.

My friend from Guyana, Alsop, had a Britisher before him who did the same thing. He told me the next day, "Jack, I was sorry for you this morning, but I am sorry for myself this afternoon."

The next symposium was on rum wastes, with much discussion taking place about reuse as fertilizer and so forth. Here, we had an Iranian professor, a poor fellow who died not long after, and he was running the show. He set up a traffic signal: red, green, and yellow. At the start, the light was green. At thirteen minutes, it turned yellow. And at fifteen minutes, it was red, and a big alarm clock sounded. An excellent idea.

I had the last paper on rum wastes. I just told what we were going to do with it and why. I didn't read any of my paper; I just used some chalk and a blackboard. My solution was to take the waste into the sewer system.

There was considerable heckling by, of all people, my clients' engineers, on the subject of corrosion the 3.9 pH waste would have on the concrete. We had built a new manhole before the waste went into it; this manhole was built by hand and plastered with hand-mixed mortar. The rungs were plain cast iron, so if any action was harmful, it would show up there, because the pipes in the ground had been made with better cement and better production. Any deterioration would be noted there, on the manhole.

I never knew why these characters all said the same thing. My only worry was that the mineral waste in the rum wastes, when mixing with the fecal wastes or domestic sewage, would not react the way I'd hoped it would, and the stabilization ponds would not produce the proper action for at least a ninety-percent removal of BOD and solids. But it worked out.

By the late sixties, of course, there was more or less a student rebellion and also Black Power demonstrations. UWI students had a small riot because of all the TV coverage; "everyone's doing it, so do we." But the Black Power issue was a more serious matter; the background of this was communistic at the base.

I had 150 local people working for me at the time, and many of them were "power" people. They used to talk to me, and I'd tell them the country would end up speaking Spanish, if they let Cuba in. If they didn't get Cuba in, the Venezuelans would take over, so either way, they would have to speak Spanish in order to be able to speak to their masters.

It started out with marches. They had Black Power anyway, but they claimed that leaders from the Prime Minister down were Afro-Saxons—black skins, white souls. These marchers did a bit of glass breaking and frightened people. The government asked my advice.

I said the first thing to do was to stop the marchers. Of course, the Roman Catholic Archbishop was with them, too. The Jamaicans were having troubles, too, and the Prime Minister thought they had troubles because they'd kept it bottled up. Of course, the Jamaicans are much more warlike than the Trinidadians.

My driver said Trinidadians "can't stand no pressure." The Prime Minister said, "I know my people. They won't rise up."

But he didn't know they already had a tree picked out where they were going to hang him. I always carried a pistol. I had to get the Prime Minister's permission, because I lived just across the street from him. I suppose he knew my weapon would be used in his defense anyhow. It was not to defend myself against a mob, of course.

About this time, Harry Kinsel, one of my oldest friends with M&E had just been made president of M&E. He and his wife came down to visit us. One day, we were in Tobago, where Harry and I were playing golf. His wife, my wife, and my son Jeff were on the beach.

Some of these marchers were around. While we were on the golf course, a few came out and broke some windows in the club house. They took clubs away from some of the caddies and threw them down, so as to protest tourism as whorism.

Later that evening, Jeff was driving, I was in front (we had a rented Volkswagen), while my wife, Harry, and his wife were in the rear seat. We met a large group, perhaps 150, mostly very young, marching down the whole road. We had to stop.

They circled around, chanting and pounding on the car. I remember my arm was on the door, and they were stroking it, saying, "White. White." They were denting the car with their pounding, and finally began rocking the car, probably preparing to turn it over, so I got out of the car. I certainly didn't want the indignity of being dumped on my head along with my family and friends.

When I got out of the car, my wife was worried. She knew I had a pistol in my pocket. She should have known I would not have drawn a weapon in the midst of an unarmed crowd. Anyway, I pushed my way. They backed up to their leaders. I said, "You gotta let us through."

The leaders said, "Get back in the car, roll up the window, and we'll get you out." I did that, and they let us through.

Helen Costello, the ambassador's wife was later to encounter this group, but outside of being scared, had no trouble. ("Oh, it's Mrs. Costello.") The trouble continued. I sent Jeff and my Maggie back to the States. I had one hired girl, Sylvia, (she did not live in), and people could not understand why I lived alone in the house. But I had a pistol and an ice-axe and good little pot hound, Pizza, if any trouble came.

We used to meet for lunch by Barclays Bank with some of the good customers. One of the luncheons took place about this time and was attended by an ex-lieutenant colonel in some Scotch regiment, a very prominent labor leader, plus Mr. Harris, Secretary of Home Affairs, and several other Barclay's customers. Talk was on tap concerning the pending crisis.

I said I had quite a few people working for me who were in with the Black Power group, and according to my best thinking, twenty-five percent of the regiment were not going to be loyal, and if ordered to protect the government, they would mutiny.

All the others said that was unthinkable. The officers were all educated at Sandhurst, and they would be loyal to the Queen and so forth. Well, the regiment mutinied.

Errol called me and asked if I could blow up the road. There was only one road along the bay, from the regiment's post to the town. I said that all the explosives (military and civilian) were in care of the regiment, and the only thing they could do was to take some old buses and trucks and block up the road with that, then have the Coast Guard patrol the adjacent bay. This was done, and the regiment was temporarily blocked.

There had been one fatality so far: a fireman was killed when a rioter threw a grenade into a fire truck. The next occurred when the Coast Guard fired on a truck at the road block; I believe one soldier was killed.

Then, the police, who remained loyal, received some arms from the U.S. and Venezuela. The Venezuelans had overflown several fighter bombers and pulled a couple of small gunboats to the dock at Port of Spain. They patrolled as well as the Coast Guard boat, whose commander was white, Capt. David Bloom. The Coast Guard remained loyal.

It always interested me that my friend, Commissioner Tim Bernard, who was head of the police, was told by Eric Williams to stay in London, where he was taking a course in law. This left his second in command, Tony May, who was white, in charge of the police.

Commissioner Bernard, who had succeeded two white men and was the first Black commissioner, did nothing about the uprising. He stayed in London, as he was ordered by the Prime Minister.

One incident rather amused me. I was with Errol, Bruce Procope (the Queen's counselor, who later prosecuted the mutineers), Clive Spencer (the Labor Leader), and a few other people at Bruce's house. There was a constable outside with an old Enfield 1898 rifle.

We heard a shot. Mrs. Procope said she did not hear anything. I went outside.

The constable said he thought he had heard some marching feet. He'd cocked his "piece." However, no one came, and he wished to un-cock it. He didn't know what to do, so he fired a round in the ground. He only had two rounds, one in the rifle and one in his pocket. I took the extra round away from him, so he would not hurt anyone, including himself.

I heard from some people that the Prime Minister and the American ambassador did not know much about military matters. "Why, they didn't know what a Karl Gustaf was."

Of course, I knew that the Swedes had such weapons (Karl Gustaf). Mostly, I heard about low-caliber 5.6 mm and so forth. "No," they said. "This is an 84mm."

I said that's a recoilless rifle; an anti-tank weapon. Why would the regiment buy such weapons? And they said, "Good question." But anyway, they had several of these.

The mutineers had dragged out one of these 84mm and were going to fire at the Coast Guard cutter, which I believe only had a 20mm Bofors gun. The 84mm would have blown the ship out of the water. Luckily, the sergeant who knew how to operate the weapon (it takes two men) threw the breach block into the bay, so the weapon could not fire.

After the fiasco, there were torturous proceedings. The brigadier, the Regimental Commander, who took no part on either side during the mutiny, sat on a rock up in the hills and collected all the soldiers who had "gone over the hill." Then, there were long trials, very expensive. Some of the troops were convicted.

Raffique Shah, a lieutenant who went to Sandhurst, too, and a foxy fellow, was the ringleader. However, the conviction amounted to nothing, because the Privy Council in London threw all the cases out, and all was the same again. As far as I know, Raffique Shah became and is still head of the Labor Congress, and of course, this is a good job.

There were hundreds of tales of Trinidad. The American ambassador, when I first arrived, was a career officer and a retired tennis player. He told my wife he would find a fine tennis coach for our children (because we had a tennis court at the house). But when I called him, he said, "Don't call me, I'll call you." He never did.

He was really annoyed with me when, at the water well turnover at the old naval station, Kamal Mohammed, the Minister of Public Utilities, introduced me to the ambassador and said, "Mr. Scheuren is the most popular American in Trinidad." The ambassador walked away in disgust.

The next Ambassador, William Costello, was not a "fat-cat" political appointment. He did not have lots of money. He had been Edward R. Murrow's man in China years before and had been White House correspondent for CBS. He had a heart attack at Lyndon Johnson's ranch and had been transferred by Lyndon Johnson's plane to Cleveland for a heart operation. When Johnson became president, he made Costello his ambassador. It turned out Bill was an excellent man, an excellent reporter, and a great correspondent.

He wrote a book, *The Truth about Nixon*, that really helped Kennedy and Johnson beat Nixon in 1960. It was really a prophetic book about all the flaws Bill saw in Nixon's character. When Nixon got elected in 1968, Costello's resignation was one of the first to be accepted by Nixon, although Kennedy's brother-in-law, Shriver, stayed a long time as ambassador to France.

Bill and Helen started building a house in Tobago. Helen finished it after Bill died. I carried his heart back to the U.S., so the surgeons could check it. It was in a sealed container made of metal and solder, and stored in formaldehyde. I had a letter with me from the Ministry of Health describing my package and had no trouble in Customs. I brought the heart to my house and, the next day, took it to M&E's office, where the mail room packed it and mailed it to Dr. Barney Crile in Cleveland, who had done the bypass operation on Bill several years before.

He wrote me that the synthetic valve he had put in was still in good shape, but that Bill's heart muscle had finally failed. Helen, a really feisty lady, continued to build her home in Tobago and lived there for several years. During the Black Power trouble, she was alone in the house part of the time, and part of the time had an artist, an American lady, with her.

Helen was an excellent painter but sort of gave it up. She called me one day and said there was some trouble in the area, so I flew over and found that her neighbor, a Black man who had some sheep (people called him Abyssinia) was walking around with a machete (called a cutlass) and muttering.

I walked up to him, saying nothing. He was just whacking around with his cutlass, so I never spoke to him. I just walked around him along Costello's property line, and he went back to his farm and made no more noise. Helen, several years later, finally sold the house and returned to the U.S.

The third Ambassador of the U.S. while we were there was J. Fife Symington. He was a cousin of the Missouri Democrat, Stewart Symington. He was a devout Republican, or of course Nixon would not have appointed him.

He and his wife, Marci, were good friends of ours. Fife and I used to play golf together, and Maggie and I played bridge with him and Marci. The only time we ever talked with or even came close to politics was when we were having our problems with Trintoplan.

It didn't look as if we were going to get the final design and construction of the national water project. Fife did suggest we talk to Maurice Stans. I did not pass this on to Mr. Eddy. After what happened later, I guess that was the right move.

During Watergate, long after Fife and Marci had left Trinidad, there was some uproar, as Marci had given a lot of money to the Nixon people so that Fife could get an ambassadorship in a better and larger country. He didn't get it, and I believe the money was given back to Marci.

When I used to go to lunch at the Union Club with Fife, or go to the golf club with him in my car or his, the Sergeant of the Special Branch (which is plainclothes, like the Secret Service) never used to ride in the car. He knew I was armed.

He was always amazed that the CIA man at the embassy (I don't remember what job he had) never knew that I was armed, and this always made me wonder how skilled the CIA people were. When I was later in Bangkok, the Intelligence Operations did not do anything to quiet my fears. Herbert Hoover said many years ago, "The West Indies is the poor house of the Western Hemisphere." The excuse of the people there is always slavery.

Dr. Eric Williams said in his book, *Capitalism and Slavery,* of which I have an autographed copy (he was a great man, Eric was, but still subjected to the prejudices built up by his treatment of the U.S., as a Black man), that we must not forget that every Black man pulled into slavery in Africa was originally sold by another Black man.

I should like to quote things and happenings in the West Indies. When I told Errol what the Prime Minister should do about the marches—stop them—Dr. Williams said, "I am a historian, and I know my people." Perhaps his handling of the violence was a real deep plot. In the end, people like Raffique Shah came into his bailiwick. But, in the long run, I don't believe he really helped these people by his rancor or about the way the U.S. treated him.

Both Williams and the Black Power people thought it was beneath their dignity to work for tourists. Time will tell how the West Indies will survive. The feeling that the officers of the regiment would have been loyal to the Queen does not count. Correct discipline and loyalty are not the same. Loyal to the Queen is not synonymous with loyal to the PNM (People's National Movement), a government party.

I took a holiday in late 1971. Before I left, I told the operators at the San Fernando plant that the digesters and the piping needed to be cleaned. The operator had called me and asked about this problem. I told him to read the operating manual. He hadn't, and the sludge had built up.

Because they had not been operating the plant correctly, the problem had grown, so they needed more people to clean out the hardened sludge. In order to get these people, they needed to put a vote to the cabinet or some such thing, or to a Board of Finance. We didn't get the vote, and the digester blew up, so the plant needed major repairs.

WR Grace & Company had a fertilizer plant there, Federation Chemicals, at Point Lisas. One of the company laws was that anyone fighting would be discharged. One day, a disgruntled employee assaulted a credit union man in the office. The man did not strike back, just held the striker until help came.

The striker was fired. He brought suit and ended up getting back pay and so forth and was reinstated in his job. The defense was that it was no fight. The victim did not hit back, so it was not a fight, and therefore the man could not be fired. This decision was upheld by the Supreme Court.

One of the biggest complaints about the sewage scheme was that it cost people to lay pipe from their property to the street to use the sewer. One of the hotel managers told me, "You are too expensive for us."

Some of the people, including a prominent doctor and head of the housing authority, complained, "You engineers have got to get this government housing built for less money. These women (one-parent families) work as domestics and get $50 a month. If the housing should cost twenty-five percent of their pay, you should build housing that can be rented for $12.50 per month."

This, of course, was impossible. When we suggested that they pay their servants more, the cry was, "You Americans pay them too much and spoil them." I heard this from the wife of one of the richest men on the island.

This woman was an American, a nurse who had nursed this rich man when he was in an auto accident in the States. I told her,

"Of course we pay our servants more money. If you can't afford to pay, I'm sorry."

She said, "What do you mean, I can't afford to pay?" and never spoke to me again.

Some ethics there were different. We had a surveyor who closed out a traverse with one long shot. When I saw it, I complained that he could not have seen from where he was one way or the other, because of the buildings in the way.

He said, "Oh, I know that survey was okay, so I just ripped over it and closed the traverse." That was his last job for us.

Then there was the local engineering firm who got a job from Julian to add to a sewage system for the Industrial Commission, which we had already been paid for and would have done the addition for free. These people hanged the Title Block and our signatures and reissued the drawings as their own. They even included the draftsman error, which we had corrected with a later drawing.

We had done a survey for sewage in Tobago and the same engineering firm tried to buy the survey notes from our survey chief to sell as his own, but no one seemed to care about that. Johnny O'Halloran's favorite expression was, "What's in it for me?"

The Black Power people, upon being asked what their manifesto was said, "We'll decide that when we get into power."

"Just now" is a favorite expression down there. It means that time is not really of the essence. It is not really "manana"; it means more that someone may do it or he may not—or it may not happen. But it notes no specific time, and he may need to be reminded later of the whole affair.

I remember looking at a problem where the pipe was to be laid near the foundation of a highway bridge. I went down into the trench and took a scaling bar and peeled off a little rock. All I could hear was, "Big boss down de hole," and the scaling bar was wrestled from my hands. It probably took the men three times the time it would have taken me to finish the job.

Once I ordered a rum at a Water Geologists meeting, and all were shocked. All wished to impress guests and so forth. They would gain prestige by serving Scotch and, of course, would drink Scotch when it was free.

Then there was the time I was arguing with Ron Bates, the Chief Technical Officer at WASA. He said, "You're coming on hard about our maintenance procedure. I want to remind you that you are only our consultant, and we do not have to do what you're saying."

I said, "I'll give you some free advice. You are paying a lot for mine. Just cancel my contract, and let me go back to New York and make more money there."

Funny thing, when I did go back to New York several years later, I did make more money, but I found that maintenance in the New York City plants was no better than in Trinidad, because of union problems.

In 1971, I wrote that work ethics, and ethics in general, may seem at variance with the memory of the U.S., but I had been away so long from my own country that perhaps the U.S. ethics and work ethics may now resemble that of the Caribbean's.

I got a telephone call one day from a woman in San Fernando. She said I might think that Frankie Gagadhar was doing work for us in San Fernando, but he was not doing the work. He was spending all day humbugging with her daughter.

She said, "I don't like that, because he's married."

When Frankie was faced with this complaint, he said, "That's true, I'm humbugging her daughter. But her daughter is married, too."

Then there was Desmond Smith's rodman in Tobago. The man did not show up one day. Desmond saw him walking toward the beach.

"What happened?" he asked.

"Oh, my friend just got a nice big dolphin. He gonna gi' me de head. I got some green figs (small bananas), and I cook it up. And man, I can go for dat."

He didn't go to work until the next day.

One of our duties in Trinidad was watching the fish in the treatment ponds. The rivers had been stocked with tilapia. Tilapia is an African perch, a mouthbrooding and very prolific fish.

Some of them got up waterways into the ponds. The oxygen content in the ponds was always close to that needed by fish. When the skies were cloudy, and there was little wind the oxygen level would drop, the fish would die and float toward the embankments, where the Corbeaux vultures would clean them up.

We did not need the fish in the ponds, but they did no harm, because the Corbeaux kept the ponds clean.

♦♦♦

# CHAPTER 7

## New York

We had pretty well completed our work in Trinidad. In the meantime, M&E had been sold to Research Cottrell. Bangor Punta, a conglomerate that had bought M&E out a few years before, had worried little about our internal affairs. RC, a larger outfit, started to take a much more active hand.

When I returned from Trinidad, I found myself, along with Mr. Eddy, Mr. Paton, Mr. Livingston, Mr. Raymond and so on, i.e., about all the officers over sixty-two years old, in various stages of layoff, discharge, or retirement. Whatever the title was, it equaled unemployed.

My wife already had a real estate license, and I, of course had an engineering license, so we set up a small office in Cohasset. Our business was soon over, as Perini Corporation was doing a five-contractor joint-venture in New York, over $250 million along the Hudson River.

Charles Richardson was then the VP in charge on the Marine Division of Perini. He had taken over when my old friend, David J. White, had passed away. I went to work as a subcontract engineer on the job. I had been away from Perini for over ten years but had been in touch.

I moved to New York City, into an apartment building rented by Perini for their supervisory help, on West 82nd Street at Riverside Drive. We had a tunnel shaft on West 79th Street and a large area from 135th to 145th Street, extending 800 feet in the river, for this project.

Perini was sponsor of the joint-venture. The co-adventurers besides Perini were Brown & Root of Texas, Horn Construction of New York, Savin of Hartford, Connecticut, and McKinney Drilling of Texas. There were also subcontractors for electrical plus some of the outfall work. Spencer White and Prentiss for the slurry wall and the tie backs along the river, to act as part of the coffer dam for the sewage plant construction, and also to protect the West Side elevated highway, which bordered the site.

We should note that the design estimate was $98 million and the low bidder was $238 million, and the second bidder was $65 million higher; obviously, the engineers were not precise. Their estimate would hardly cover the cost of the pipe. The project was merely a giant foundation for the large sewage treatment plant that was to be built there.

The foundation was to be set on 2,400 caissons up to four feet in diameter; these were pipes driven into the material on the shore and out 800 feet into the river. Many were over 200 feet long. The project was set up with several drill barges; each of which had two mounted Manitowoc cranes, with 150-ton capacity. These cranes had several functions. One was to pick up and set the pipe for the caissons, then set a pile hammer over the pipe and drive it as far as it could. Then, the other crane would put an airlift inside to suck all the dirt and rocks inside.

This operation of driving and airlifting would continue until the pipe hit rock bottom. It was then driven to set into the rock. Then, the drill stem was set on the pipe and the drilling would start, with the tool like the one used for oil or gas drilling. After the hole was down to the rock to a specified depth, the airlift was used again to clean out the hole for placing of the core beam. The pipe had to be driven several times to seal into the rock, until you could get a seal that could keep the silt and sand blowing in from the bottom.

After the seals were complete and the hole cleaned out, the core beam was set in place, and the grouting, a mixture of sand, cement, and water, was pumped into the pipe along the core beam and a few feet above. This was done by a pump. I had disagreed with the procedure of pumping the tremie material. It is best to pump tremie material to a hopper on the surface and let gravity take over.

Then, an attempt would be made to clean off the top of the hardened grout and pump the concrete mix of cement, sand, water, and gravel, to fill the pipe up to grade. There was a loss of seal many times, and this effort was often a failure, so a great deal of unsatisfactory concrete would get in and create a flaw and get laitance in the caissons. Repair work would be required.

We were using rotary drills on the caisson foundation with attendant problems. However, the subcontractor (Spenser, White and Prentiss) on the slurry wall (a CPFF job for them, although hard money for the joint venture), were using churn-type drills—percussion type, rather than rotary. These had the advantage of being a bit slower but did not cause a loss of seal.

I wrote a report about this to the director of the joint venture, recommending they use churn drills on the rest of the drilling. This was ignored for quite a while. It was not until I had left the project to return to M&E on a large sewage project, a year later, that the joint venture finally rigged up a couple of barges with churn drills and found these most satisfactory.

This subcontractor (Spenser, White and Prentiss) had more experience in foundation work in slurry wall construction, but they were CPFF, and I had a really tough job keeping them from cheating on the job. They would bring down a rig to the job and start charging rental for it and then spend days of labor and costly spare parts to get it going; all of which had to be paid for by the joint venture. The superintendent for Spenser used to give me lots of trouble. I would cut out paying for these mistakes, and he used to say he had twenty-five years' experience in cost-plus work. The only trouble was, he had a cost-plus job, and the joint-venture was a hard money job.

My duties called for me to check with the office people on the reports of the various foremen and superintendents, to see that they had charged the correct item for their time and material. The cost system was too elaborate for these dock builder foremen and dredge operators to keep that kind of track. That was not too much of a job for me, but my co-adventurers used the same type bills, which I had to check. These companies would do as Spenser did: bring a crane on the job that needed boom repairs and without any cable clamps to get the equipment ready for work, and then try to charge the joint-venture for a regular rental rate while it was being repaired.

I remember our chief accountant telling me that "this damned engineer (as these co-adventurers called me) was giving them plenty of trouble by not okaying their bills." While I didn't okay it, no one could get paid unless the bill was marked okay by JJS.

The job was behind schedule. The estimate, of course, had been too low, and there was lots of trouble with their methods. The City of New York did not have enough money, and things dragged on. The contract finally got done for the foundation, but the rest of the treatment plant was not done yet—thirteen years later. This was an interesting exercise, and I learned a lot.

When I left there in early in 1973 to go back to M&E, Albert Perini told me he was sorry to see me go. He'd realized I was overqualified for the work they had me doing. Louie, his father, died in early 1972. Albert left the company a couple of years later, because there could only be one person as president, and his brother David won out. Albert is now with a big international rigging company in San Francisco.

M&E made me an offer I couldn't refuse, so I went off to manage the construction of a large sewage treatment plant on the other side of New York City. By then, I had moved my wife and my daughter, Margot, down to New York. We had a townhouse in Port Washington. I had a Perini car assigned to me, but the traffic was such that I never used it and took the Long Island Railroad, which worked out better.

When I started work at Rockaway for M&E, M&E furnished me a new Chevrolet Impala every year for my use. This job had been designed by M&E several years before, and the contract had been delayed because of lack of finances. In the meantime, some of the facilities had been set up and the design and procedures to be used during the construction of other stages had been worn out and were in a state of disrepair. We had to redo the schedule of operations. Actually, this was a boon, because I found we could fill it with different sequence—we actually had to do better and save money and time.

Before I left the Perini job at North River, I took one day to meet at M&E offices with some of the Dept. of Water Resources (DWR) engineers and drove out to Rockaway. At this time, I had not seen the job and had not looked at the specification and drawings. The Rockaway plant was in a state of disrepair, and part of the operation tanks were completely useless from lack of maintenance. I found the final tanks were inoperative.

When I pointed this out, the City engineers were shocked. I said I could see the motors and the chains were not moving. They opened the hatches and found the tanks were, of course, empty. They were shocked because the plan to keep the existing plant going during the construction of the new addition included the use of these final tanks for a certain period.

When the drawings had been made by M&E, the tanks were operating. No one from M&E, nor I, could surmise that DWR had looked at the plant for several years and had let the contract without a field check. As it turned out, however, this enabled me to revise the schedule of operations, which I did with the help of the Project Engineer in Boston, Carmine DeFilippo, whose father, incidentally, had been one of the caisson workers on my job with the Gow Company in 1929-30 and the Kaiser, Nick Fiori was Carmine's uncle.

The City people complained about my rescheduling, but I believed I was in charge, and I had to get the job done. The contractors also complained but only said they reserved the right to make later claims. But they never had to do that, because they actually saved money by the revision. Later, the Commissioner, Marty Lang, had joined up with Camp Dresser & McKee, and he used this change in sequence to praise me and put on my resume to sell me to the Water Commission in Jamaica.

In New York City and New York State, there is really no general contractor unless it is the owner. The job is split up between five prime contractors: structures and equipment, plumbing, electrical, heating and ventilation, and instrumentation. The structures and equipment are the first in line. They have to do work so the others have walls, floors, and ceilings to append their work to. This, of course, protects these people, who would have been subcontractors for a general contractor who, under some circumstances, would then cut their prices. But this makes the contract more difficult to administer.

Because the engineer had been unfamiliar with the building commissioner's regulations and did not know M&E had to get the building permit, there was lots of trouble at the beginning. All the drawings had to be approved by the Building Department before any work could be done.

We were able to get a partial permit so that excavation could be started—not more than ten days, ten-feet deep, and fences put up and so forth. Then, the fun started. All sorts of roadblocks were thrown up. Many revisions had to be made, and I found that the building and equipment contractor had an expediter who cared for such things as getting permits.

I visited this expeditor. He had a nice office with draftsmen and so forth and would make any minor changes to get the drawings approved. I made a deal with him for not much money. It was a lump sum plus an hourly rate for any drafting work he might have to do. I knew that part of his fee would go to paying $25 or so, here and there, to various clerks and inspectors in the Building Department. Later, this man was indicted but never convicted.

Things started to get better. Then, the Boston office began to think this was bribery and told me to cancel Pete's contract. Pete took it okay, and M&E sent down a group of architects and engineers to get the drawings approved. This was, of course, an expensive procedure, but more than that, when this group arrived at the Building Department, the drawings somehow could not be found. Lots of money had been spent, and we would have to start all over again.

I called up my boss, Al Thomas, and asked him if he had had enough. We were paying this expeditor for the service he was supplying. If he spent a few dollars on paying off a few people, it was no problem to us. We charged his bills out to expenses, and neither could we be charged for bribery, nor could the IRS complain about what we'd paid. So, they capitulated, and the drawings were found forthwith and the job proceeded. I am sure, without Pete, even if there were no payoff, the inefficiency of the Building Department would have soon delayed us and cost of lot of money. (*TIP* is supposed to mean To Insure Promptness.)

The next problem with my main office was over the hiring of my daughter, Margot, to manage the office. She had been working for Blue Cross-Blue Shield as a system analyst, but the atmosphere in Boston, with robberies and so forth, was such that she moved in with us at Port Washington and hoped to get a job in New York City for Blue Cross. The liaison engineer for the City met her and asked me why we didn't hire her for office manager. But Boston again got in on the act and told Al Thomas I should let her go—nepotism. I told Al, if she goes, get someone to take my place. Anyway, the case was closed, and she did a splendid job.

Under our contract, the building contractor, Fortunato Contracting, furnished us a large, double-trailer type office, and it was soon manned with Desmond Smith, whom I had imported from Trinidad; Margot; an electrical engineer from our New York office; a structural man; a materials man; and a female engineer, Marilyn Roberti; plus an assistant for her. Marilyn used to climb around the forms like a monkey. She ruled over the concrete placing like an empress. If the concrete was too wet, back it went. If the steel wasn't in the right place, it got changed. Of course, Desmond was his fine self and fell into place as if he had lived in New York City all of his life.

One of the good things about managing this difficult job was the Structures & Equipment Contractor's superintendent, Don Anspacher. He had started as a laborer in the business and held all jobs to General Super. He was excellent and had two people with him, Jack Mahoney and Pete Berotta. Jack was on the structures part, and Pete was on the mechanical part, and the job went well.

The other contractors gave us a lot of trouble. The electrical contractor was a pleasant fellow, but his problems with the Local 3 of the IBEW kept him in line with the union. But no matter what you say about Local 3, the contractor can't even set foot on the site. Each of the workers are good soldiers of the union, and many of the union rules on safety, etc. are valuable. And all of the plumbers weren't really trained in safety procedures. The plumbers and the HVAC were nitpickers and tended to co-inspect some other jobs and so forth.

The instrumentation contractor, Honeywell, had no experience in this type of work. One of their instruments was a large steel casting for a Venturi tube. They held up the job because they thought it was the duty of the structural contractor to put a large crane on the job, to set it for them. They finally made a proper deal, and Anspacher had it set for them, for a fee of course.

This, of course, was an affirmative action job, and we had to show our use of minorities. However, when the man came from the Labor people of the U.S. government, he was an old heavyweight boxer, Billy Martin, whom I had met years before, and he was Black.

He came into our office, chatted with me, and said, "Jack, sure no discrimination here."

I had two women, Margot and our concrete inspector, plus a mixed Black/Chinese, South Korean, Puerto Rican, and a Black male typist. No problem with minorities.

Margot ran the business part of the job with an iron hand, but also drove me to work each day, which was a real project. I had taken the apartment in Port Washington because it was handy to the Long Island Railroad into Penn Station, where I took the 7th Avenue subway up to 125th Street or 132nd Street, to work for

Perini. From Port Washington to Rockaway was a rough drive across Long Island, but Margot handled it very well.

The total bid price for the work at Rockaway was over $40 million and included some allowance items for welding inspection, temporary heating, and temporary electric power. Because we had changed the sequence of the job, we did not need so much electric and heating, which saved a considerable portion of the allowance. The welding inspection was cut because I did some myself, and the pile testing was done by my old pile-driving man, Harry Corfield, who had been with us with his brother, Joe at North River.

There was a subcontractor on the work on the refurbishing of the existing dock, the laying of the outfall, driving piles for the outfall, and the piles for the foundation of the sludge storage tank, where the sludge was stored, preparatory to being taken aboard a sludge vessel and dumped out at sea. The subcontractor was an old firm with a famous old name: Spearin, Preston and Burrows (SP&B). All of the old firm was long gone, and the present management was headed by an in-law of the Burrows family. The general manager was a retired Navy admiral, and the job manager, Ernie Swenson, was a former U.S. civil service employee, who had been working on roads and streets at Governor's Island.

The contract called for them to furnish a power boat for our inspector to use. This was a little frosting on the cake, and at North River, Perini had offered a credit not to furnish the boat. Here, the admiral was too stupid to offer a credit. He told us that he would furnish a canoe, and I told him to try to get paid.

The specifications called for piles to be at least five-foot penetration and to submit a load test. SP&B kept cutting out the piles at five feet under the sludge tank. I suggested he jet the piles down and get more footage. He was getting paid by the foot.

He would not take my advice and ended up with more piles to haul to the dump than he had in the ground. They were creosoted, so it was a long haul to a certified waste dump in New Jersey.

The outfall was to be set on pile bents. There was some excavation to be done by dredging to about two feet below the pile tops, then the piles had to be driven and the bottom leveled up with gravel, and the pile caps placed by divers and the pipe laid on top. Then, the whole thing backfilled with gravel and covered with rock on top. SP&B seemed to have some customer for the dredge spoil, because they dredged too deep and hauled the river mud and silt away. Then, when they had driven the piles, they wanted to put the caps on, then the pipe, and backfill by clam-shelling the material from outside the area, casting it over the outfall, and dumping it over the pipe. They put a bunch of piles into the bay and did not have proper survey control. They had to drive a lot of extra ones. So, they bought more piles than they needed.

Before they went to lay the pipe, we found that the piles were exposed in the trench, sometimes nine to twelve feet deep, and they proposed to lay the pipe that way. The divers had to nail cleats on the piles to get up there to lay the pipe, to put the caps on. This was dangerous, of course, and besides, the pipe was specified to be set and backfilled with gravel.

There were two meetings, and I had to defend my position before SP&B and also the DWR people. Then, they made two mistakes: 1) they tried to bring political pressure on the director of the DWR, Joe Miller, and 2) they brought a lawyer to the second meeting.

When I found that out, I got up and said, "I have not brought counsel, and until I have, I won't participate."

All the DWR people got up and left too, and the meeting was over. Finally, they got some gravel and filled in the excavated trench. After they laid the pipe, they brought in more gravel to finish the job. It really cost them a lot of money to sell that dredged material.

Then, we had the welding of the sludge tank. For this, I hired a man from Haller Associates, an old company specializing in materials engineering. This welding inspector, Frankie, did a good job.

Then, they hired another contractor for the tank for sludge gas, methane, which was used to heat up sludge, and also to help heat the buildings. This welding contractor did not have his men qualified, and we would not accept his work.

These people had built a couple of other tanks for the DWR. One place, they had checked the specified percentage of the welds, and repaired any failure, but did not check the others. This is not how you do it.

When we finally left the job, we had not signed off on this tank. I refused to put my inspectors on it. But anyway, Herb Tschudi of DWR, finally signed it off, himself. I hope it held together.

*North River, New York City 1*

*North River, New York City-2*

*North River, New York City-3*

*North River, New York City-4*

*Rockaway, New York*

*Rockaway Next Door Idea of Sheeting*

*Rockaway Pour Space for Chlorine Contact Tank*

*Rockaway Scaffolding*

*Rockaway Primary Sed Tanks*

♦ ♦ ♦

# CHAPTER 8

## Thailand

In early 1976, M&E called me about going to a job in Thailand. They'd called me the year before to ask if I knew any good tunnel men. They had recommended a good man, Jerry, but he was retired and said there would probably be Japanese contractors there, and he did not like their safety methods. Besides, his wife had a good job, and she did not want to leave New York.

When they described the job to me at that time, I questioned the use of reinforced concrete lining in a soft-ground pressure tunnel, and said it should be lined with steel plate. However, I was not asked then to go myself.

In March 1976, M&E started approaching me to go to Bangkok, but I couldn't come to agreement with a price and perks so I retired in June 1976, bought a 1976 Impala and a membership in a golf club and went back to Hingham. They gave me a party in Boston, and George Tozer told me then that he still wanted me to go to Bangkok. Anyway, finally, in October 1976, I got the deal I wanted and went to Bangkok, Thailand.

The original design of the tunnels, treatment plants, and pumping stations in Bangkok had been done by CDM with Jenny as their tunnel consultant. They still did not have a lining of steel plate. M&E had managed to get the rest of the work that had been partially designed by CDM. In fact, M&E had been censured by the American Society of Civil Engineers (ASCE), and George Tozer and Frank Sunn, then president of M&E, were quickly kicked out of the

Society. I always thought they should have been kicked out, but not for ethics; just on the basis of ineptness for taking the job at all.

The basis was this: CDM had a Thai subcontractor, TEC (Thai Engineering Company) headed by Procob and Rangsi, both of whom had master's degrees in structural engineering from the University of Illinois, which is an excellent school. CDM had not been satisfied with their performance and intended to dump them and get another Thai subcontractor on the remaining work.

CDM's Bob Marini, one of the senior vice presidents, had been at school with Pracha, a Thai who was high ranking in the Waterworks in Bangkok. He was an enemy of TEC, which was backed by Birabandthu, who was a classmate of Rangsi and also prominent in the Waterworks.

TEC called Tozer and Sunn, and they went there and got the job away from CDM. Later, Birabandthu showed me some maps. CDM and Pracha had figured out a way to change the location of one of the intake canals to the new system, which would require a new road to be built, which would open up to development some land owned by Pracha or for which he had an option on. True or not, I cannot say, but there were bitter feelings.

Anyway, M&E got the job and sent Bill Roper, a retired major general from the Corps of Engineers, to manage it. Bill had graduated from MIT in 1938, was in ROTC, and was in the Army early on, building B29 bases in the Pacific.

After the war, he was the District Engineer Division Engineer in various Army jobs. Bill had been retired from the Army because of illness, emphysema. Unknown to me, a complaint had been issued, like an indictment, against M&E by the client, many months before. It said something like, "You have not done your job," and we were almost kicked out of the country. M&E figured I could do something about it. I did, and the complaint was withdrawn.

The problem of secondary lining still persisted. One of the Japanese contractors offered to line his section of the tunnel with steel but proposed an epoxy coating on the steel. If they did that, M&E said they would have to use cement mortar coating on the steel. Then, M&E hired a subconsultant, Mott, Hay & Anderson, a prestigious British firm, who checked into the matter. Their report added more rebar to the reinforced concrete secondary lining.

Jenny was asked, with CDM's permission granted, to comment, and his reply was that they "unofficially stood by the original design." I don't know how it could be "unofficially," if they had already stamped it with their stamp.

Then, the client hired another firm, half Thai and half British, who produced a new design, adding more rebar. When these people came to my office, I asked them why they didn't take all the reinforcing steel and roll it into steel plate.

Anyway, I would not stay to participate in this reinforced concrete secondary lining and left before it was started. The contractor on the river section did have the grace to use steel lining with cement mortar coating. The others used the rebar and had leak problems. The British end of the joint venture had better ground in their section, and their leaks were minimized.

I saw a clipping from a Bangkok paper last year that told of a shutdown for repair, and someone said, "Here we are with a water shortage in the midst of the rainy season."

The contractors all used compressed air when I was there. But one was planning to use the slurry shield for his last section, and this one had plenty of leaks in it. One of the contractors could not keep his shield in line—it got to be one meter off, and his engineer faked the papers to look as if it was okay.

Our resident in the section was fired, and the Japanese man went home, also. Then, we had to realign the tunnel. We had all British tunnel engineers. This was mostly because of the tax provisions in the U.S. and the UK about overseas work. And also, the British had been building more, so they had more experience on soft-ground tunnels than we had in the U.S.

It was very attractive for the British engineers, because they had no tax to pay, U.S. or UK, and the Americans had to pay high income taxes. Of course, there have always been more soft-ground tunnels, per capita, in the UK than in the U.S.

One of the Japanese contractors, Bayashi-Gumi, was way behind schedule on the job at Lumpini Park. This was a reservoir and pumping station. The piles that were the foundation of the filtered water reservoir had been found wanting, and redesign was required. This the contractor had to pay for, but it was done by our consultants, TEC. M&E had no structural design people there, and TEC furnished this service.

TEC took an awfully long time doing this redesign and delayed the job. However, the pumping station, which had no reason for delay, was also behind. When it came time to assess liquidated damages, the Japanese claimed, because of the delay on this redesign (of course, the cause of the redesign was their error), they should be relieved of paying liquidated damages on the whole project.

One reason for the liquidated damages here was that Lumpini Park was a valuable recreation area in Bangkok, and it was not desirable to keep it away from the public for too long. This question ended up in arbitration—very interesting.

The Waterworks had a lawyer, a government employee. The Japanese had a Bengali, educated at Lincoln Inns in London. I had my Thai subcontractors, who never said a word during the whole discussion, and the Waterworks had their people. The language of these contracts is always in English, but the contracts also say you must abide by the laws and customs of the country. The laws and customs were in Thai.

Anyway, they finally recommended that the liquidated damages should be applied to the pumping station, and that the time for completion of the filtered water reservoir be extended by two-thirds of the time that TEC had used up in the redesign. This was accepted, and all went well. Almost, I guess. I got a personal letter from the contractor's local representative, written in English, yet criticized me by saying, "You are illiterate." I am illiterate in Thai, Japanese, and Bengali.

There was another incident that reinforced my feelings about the insensitivity of the various home offices to the problems of overseas management. I asked the president of M&E International why he had sent a certain pair of people from Boston to Thailand. These two had been messing up all the time and were thorns in my flesh, and Bert said, "We wanted to get them out of the Boston office." Thanks a lot!

I'm afraid this type of deal is not only an M&E problem. I found CDM doing the same thing in Jamaica. And Hazen & Sawyer in Yemen. It is pleasant to get people on some steady payroll that you were making a profit on. I found that, in Egypt, both M&E and CH2M Hill did the same thing, and so did Hazen & Sawyer. In the Arctic, we did better, and there was very little of this going on. Everyone has to be alert, because the home offices like to fill positions, and either don't care or personnel people don't understand the problem. If you do not watch out, they will send incompetent people just to fill the jobs.

We arrived in Bangkok the night before a coup. After a few people got killed, we were confined to quarters in the hotel for a while. There were two other coups during the twenty months we were there. One of them I call the "cuckoo coup"!

I had heard rumors of this one from some of my Thai employees, and so had some inkling. At that time, Lincoln Bloomfield, professor of political science at MIT, and his wife were visiting Bangkok. He was on sabbatical in Geneva with the United Nations. I had never met him before, but his wife and my wife belonged to the same women's group in Cohasset. She knew I was away in the Arctic and Antarctic much of the time, and I had never caught up with them.

There was a dinner at the MIT Club in Bangkok, and they were at the function. Maggie and Ilea Bloomfield were laughing together... "Imagine seeing you here…" I remember that each one got up in the meeting and described who they were and what they were doing.

A man from Australia told how he had a PhD from MIT, but he was not sure. In fact, he was sure, that he couldn't get in as a freshman to get a bachelor of science degree. This was the way it is.

When my son, Fritz, married a girl several years back, he told me that she had a master's degree from MIT, and I asked him if she had a bachelor's degree from there also. He said she had, and then later she told Fritz she knew I would ask that question. Anyway, in my turn, I said I was a consultant for the Waterworks and my wife was a consultant to me.

Bloomfield was in another meeting a few days later. This was under the auspices of the American Chamber of Commerce of Thailand. Jack Scott was the director. He invited me as the representative of an American firm. The other American firm representatives were Cliff Bevens, a local manager for Goodyear, who lived in the same apartment house as we did, the Royal Mansions, known to some people as the American ghetto—pretty fancy ghetto! The manager of 3M was also there, and the local manager of the Bank of America. Besides Bloomfield, there was the head of the USIS. So, it was a small group, mainly to brief Bloomfield on the local scene. The "cuckoo coup" was about to cook.

The day of our meeting was Thursday, and everyone knew that the generals and admirals would never have a coup on the weekend, because it would interfere with their golf. Well, Bloomfield and the USIS man came late. They had been at a lunch at the embassy with the ambassador and General Kriensak, the Supreme Commander.

The USIS man said they were late because the general kept hanging around. USIS said he usually left like a shot after lunch. I wondered about that, and I told Jack it sounded like a Sherlock Holmes story about the dog that didn't bark in the night. I called my Thai partner, and he said that he hadn't heard anything and if it was going to happen, it wouldn't happen until after the weekend anyway, after the golf.

The discussion we had was all about local business, and we talked about the lack of patent law in Thailand. A Thai leader had once said, "The Thais are not inventive. We don't need any patent protection, and we can copy what we need." Bloomfield asked Scott what was going to be done about this, but Jack said this was hopeless.

I said I had been in a shop a few days before, where they were copying a Japanese truck. Anyway, on Saturday morning, about 9:00, when my wife and I were at Jim Thompson's store, buying some Thai silk pillowcases, the word came, and they had to shut down.

One of the generals was trying to march his army downtown to take over. So, we all went home. The other generals, including Kriensak, held their troops together, and it all came to naught. The rebellious general killed the commander of the nearest Army Corps. The man would not give the order to his troops, so the general started shooting. He shot the man's feet; but he still would not comply. He shot his knees, then shot a little higher and higher, and he still wouldn't comply. So, he finally shot him in the head and killed him.

The rebellious general was taken into custody. The other generals did not wish to try him or kill him, so they took him to Taiwan. The Taiwanese would not accept him; so finally, he was taken back to Thailand and executed.

My opinion of the U.S. security information gathering was, besides the fault of USIS having no suspicion of the general's transactions, the fact that the Naval attaché in Bangkok, Naval Intelligence Commander, was playing bridge on Saturday and did not hear of the coup until late afternoon, when his maid told him she heard it on Thai radio.

We were very fortunate in our travels around the world. We were in the Black Power uprisings in Trinidad and Tobago, went through three coups in Thailand, were in Istanbul when the Maoists were fighting the Stalinists, and we were confined to quarters at Robert College. When we were in Alexandria, we had meetings about plans to evacuate, if the Israeli business got tougher.

We were in Rome at the first audience the Pope gave after he recovered from his attempted assassination, and we returned to

Cairo in the first civilian plane into the airport after Anwar Sadat was assassinated.

Because of the last Jamaican elections, leading to 800 deaths, I worried about the 1983 elections, so I'd sent Maggie and Christen home. But then, the Manley group decided not to contest, and there was no fighting the elections.

In the U.S., there was always talk about aliens, legal and illegal. In other countries, the control of aliens is very tight. In Jamaica, Maggie and I had to get not only visas on our passports, but alien registration cards, including photographs, which we had to carry with us and show to any constable or any immigration officer at any time.

When Christen came to visit us and go to school there, we had a great deal of trouble. First, they would not let her enter the country, because her air ticket did not include an outgoing ticket, so I bought a ticket for her. Then, the immigration people said she would have to go back to the U.S., get a visa, and return, because, as a tourist, she could stay for six months with just a passport and an outgoing ticket; but because she was going to school, she must have a visa for the six months.

The immigration person was very nasty. We were doing work for the government, of course, and I talked to the Minister of Public Utilities. This permanent secretary had to write a letter to the Minister of Security and Immigration to get her a proper visa without going back to the U.S. She was only eleven years old and was going to a private school. I had to furnish a financial statement and all that.

So, although the Jamaicans are always complaining about their treatment in the U.S., the U.S. Consulate Service, on visas, there seems to be lots of them getting into the U.S. Perhaps they thought the little eleven-year-old girl was going to be a bomb thrower!

We were in Trinidad for over eight years, and I was under contract with the government there. However, for about six years, every time I left Trinidad and Tobago, I had to go through the routine to getting a new visa each time. Finally, I got a multiple-entry visa, so I didn't have to go through the red tape each time I traveled.

In Bangkok, it was a similar matter. Each time we left, we had to go through the routine of getting a new entrance permit. This meant standing in line in the sun for hours. I managed to get my wife out of this by telling the immigration people that she was ill.

In Egypt, we were better treated. We had a partner, an Egyptian firm, and they had arranged for a colonel, retired from the police, to meet us and also escort us on our way out. He was known as Captain Youssef, and he would take our permits, stubs, passports, and tickets and then leave us in the lounge. After a short time, he would return with all our luggage and our passports all checked, and then drive us to our hotel.

In Alexandria, Salah, the office manager for our local partner, would take our passports and get work permits and visas all set. A very civilized procedure. I do remember an earlier trip to Cairo from Bangkok, when, although we had a visa we got at the Egyptian Embassy in Bangkok, we had to fill out all the same papers and get the same visa we had received in Bangkok.

In 1976, when I went to Nigeria, I had a one-month visa from the Nigerian Embassy in Washington. I really only intended to stay two weeks. I presented my passport with the one-month visa, and the officer there wanted to know if I was going to work.

I said, "No, I was only looking for business," which I was. He said he didn't believe I needed a month and stamped over the month, telling me to report in fourteen days. I left Nigeria before that time. Nigeria is very fussy about people coming in to work, or they were at that time.

There was a young American on the same plane with me, who was really coming over here to work. He had been told by his employer to say he wasn't coming to work, and then they would get his visa straightened out, after he was there a couple of weeks. However, he managed to make the Nigerian understand that he was coming in to work, and it took his company a few days to get him out of jail, where he had been promptly deposited. So, all things being equal, the U.S. is not too hard on immigrants.

My two émigrés from Trinidad and Tobago, Desmond Smith and Leslie Donald, made it okay to the U.S. without too much of a hassle.

To get back to Bangkok, we found, of course, a different culture. Both Buddhists and Muslims there have a society that is primarily for men, although there are many women who hold high positions. The men, if they can afford it, and sometimes not, have minor wives. There was a word for fun, "*sanuk,*" which means pleasure is mostly for male Thais.

The Thai are good workers. They have a large number of women laborers who mix concrete, cut, bend, and set reinforcing rod, and are generally the backbone of the construction industry. These women usually wear scarves over their faces to cut down exposure to the sun.

There are ethnic Chinese there, similar in appearance, except that the Chinese have lighter complexions, and the Thais seem to have poorer skin, acne, etc. These ethnic Chinese are usually Buddhists and a little different from the Thai Buddhists. The Thai cremate their dead, whereas the Chinese bury them. The Chinese Thai are very prominent in business. But the Chinese are still not considered real citizens and have to register every year.

We went to several weddings and other celebrations. The first one was our landlady's daughter, the woman who ran the Royal Mansions. She was married to a doctor of meteorology, a very pleasant man whom I used to see at various scientific symposiums. His wife, the landlady, was known to the tenants as the "dragon lady." She was very hard on the help and on the tenants.

The wedding of the daughter was to a man named Wallop. Both families had a lot of money. In Thailand, such weddings take place at a time determined by an astrologer. The guests are not usually there, but come to the reception. The Wallop affair was very posh. All gifts are brought into a sort of checkroom in the reception hall.

Each is logged in by donor's name and placed on shelves. Then, the donors are given a small present. In the Wallop affair, each of us got a key ring with the date and names of the bride and groom engraved on the tag. There was no formal note of thanks.

Another wedding was for my secretary. She married a young lieutenant in the army. This celebration was with drinks and canapés only, no dinner, as at the Wallop affair. Here, their gift was a sprig of flowers.

My driver, Tanyoi, married a girl whose father owned a large farm way out in the country. The first step here was the purchase of the bride. Tanyoi borrowed $600 from me, well in advance, and paid me back out of his weekly salary. This he had to pay the girl's father, because he was taking a good field hand away.

The wedding was about 150 miles away from Bangkok, and Mr. Chalong, our Thai office manager, went with me. The wedding was supposed to be at 8:30, and we arrived at that time. But the wedding was all over, because the monk had another wedding to perform, so he had to do this one at 6:00 that morning. The bride was a fine-looking, strapping girl. She wore a brocaded yellow gown but no shoes. Her father had several new, shiny pieces of U.S.-made farm equipment in his yard and also a couple of buffalo who gazed quietly at the wedding group.

The house was built on stilts, and a rough ladder led to the first floor. Mats were spread, and plates of food were on the mats. The drink was 7-Up and Mekong whiskey, a sort of moonshine. (Actually, it was bootleg but made in Thailand, but they claimed it was made under the auspices of the King.) Here the gift from the bridal pair was a small jar of Tiger Balm.

One of the bride's brothers was engaged in eating his, and when I asked about it, was told he always does it. The boy was about seven years old. Of course, I knew that people ate it. Rangsi used to eat a bit off the end of a toothpick to help his stomach, he used to tell me.

The bride's father was well along with the Mekong whiskey and spent some time trying to ask me a question. Chalong translated. The old man wanted to know if Thai women were as good in bed as American women! I could only say I had no experience with Thai women. But Chalong, who had worked for an airline company in the U.S., finally told the father of the bride that both Thai and U.S. women were equally good.

This Tiger Balm is a sort of white paste or jelly containing camphor, among other things. It was developed by Chinese brothers, the Awhaw brothers. They had made a huge fortune and now controlled a huge banking company. They had showplaces in both Hong Kong and Singapore, each called Tiger Balm Gardens. These places have gardens festooned with flowers and shrubs, and also clay, concrete, and plastic models of all sorts of animals and museums had all sorts of colored jade carved in beautiful forms.

The Balm, besides being handy for stomachs, is great for muscle relief, for scratches, and bites and sores. Of course, we have a similar product in the U.S. called Bag Balm. This is used for treatment of irritations of cow's teats, but is excellent for cuts, strains, bites and so forth. An old blaster I knew demonstrated it to me years ago, and my father used it to cure athlete's foot at one time. When I ran a rusty nail into my hand, Bag Balm cleaned it up very soon.

Another wedding we went to was the daughter of General Yukontorn of the Supreme Command. He had been with Thai troops in Korea, along with the U.S. troops, and had been awarded the Commander of the Legion of Merit, only given to foreign officers. Here, the gifts were small, gold-plated pillboxes, one of which I still have. These were engraved with the date and initials of the bride and groom. This was quite an affair, all kinds of food and booze. There were very few non-Thais there. Four Americans, Maggie and I and our friends from Colgate, the Nieces. Ed Niece had Yukontorn's daughter as a secretary. The only other non-Thais were the Australian military attaché and his wife.

The general wore a khaki uniform. The bridegroom wore a white, dress uniform. The place was full of Thai military officers, all in civilian clothes. They were a rough looking group. I remember telling Niece I wouldn't want to meet any of them in a dark alley. They were all pleasant, however, after their first suspicions wore off.

Besides weddings, we were at several other functions, like funerals. Maggie did not go to the funerals. She didn't have a long black dress. My uniform was a business suit with a black tie and black brassard. I shall describe one.

This cremation and funeral was of a brother of one of our Thai partners, who had died eighteen months before. The poor get cremated right away, but the wealthy get embalmed and stored away until a book is written about their lives, which includes photographs (baby photographs, golf-playing, graduation pictures) and the story of the man's life, along with poetry and prose. Here, we were each given a sprig of pine and a flower.

After we sat for a while, not crossing our legs (crossing your legs is strictly taboo in Thailand, especially if you show the sole of your foot), when the monks finished their prayers, we were called up. Each of us came up with our little bit of firewood and placed it under the coffin. The coffin was then wheeled into the furnace, I guess you'd call it, and by use of natural gas for fuel, the body was cremated. Only a few of the close friends of the family stayed there then, and we others left.

The family and the monks would rake around the ashes and find a piece or two of bone. These pieces of bone would then be cemented into the tiles of the floor of the temple, and the rest placed in an urn. Nothing is ever done without proper ceremony.

When the Japanese contractors, who were Buddhist, were going to start building their part of the work, the date and time were set by the astrologers. This particular ceremony was set at 0845 one day. There were nine monks there. They sat in a line, with a long cord stretched between them. This cord they handled and chanted over during the ceremony. The next thing was to feed the monks. Of course, they had not eaten anything that day yet, so each of the monks were given food, and the rest of us were served quite a repast. Then, the construction started.

There was a celebration for the end of the Buddhist Lent. There, the monks are given new robes. I was the only one of our people invited to this function. I had, of course, removed my shoes at the wat entrance (wat is the word for temple). I sat down with the rest of the laymen in front of the altar. I was the only non-Thai there. Everyone moved around barefoot and on their knees around the altar, and the monks, in their saffron robes, sat in several rows. There were a few female monks there, Buddhist women in white robes, also with shaved heads.

Mr. Krachok, the director of the Waterworks, was acting as the representative of the King in this ceremony, so he had his shoes on. I didn't know what was going on. I just sat quietly with my feet flat on the floor until the ceremony was over and the robes distributed to the monks. I do not remember if they gave any robes to the nuns.

This ceremony had odd results. Pracha, the friend of CDM, was there, and when we left the temple, he told me that he was surprised that, as a Christian, I would take part in the Buddhist ceremony. I told him I followed the Unitarian beliefs that Buddha was a prophet, as well as Jesus and Mohammed. He was quite impressed and, thereafter, was very respectful and used to come to ask my advice about maintenance of his stuff.

Pracha had some deal going on with CDM to relocate a Klong road to increase the value of some land he wished to develop. Birasandtho squelched the plan. Later, Pracha got back in power. Pracob and Birasandthu went to jail, I believe, for a time.

We had to take part in a tree-planting ceremony along the Klong near our office. I had to plant some trees; I hope they grew and prospered. I didn't understand the Thai language, so I did the best I could. I guess I made no bad mistakes.

Buddhism teaches reincarnation. If you are decent during your life on Earth, you will be reincarnated into a person of higher rank. If you have not been a good person, you may come back as a dog. There was one little dog who lay around our office out at the plant, all mangy and sick. He should have been put away, but no one wanted to do it. He might be the reincarnation of somebody's uncle.

There was a meeting once at the American Chamber of Commerce with reporters and dignitaries on a discussion of politics and economics. I remember an economist saying that the U.S. dollar might drop in value to almost 200 Japanese yen. All were horrified. It dropped lower than that. There was and is much turmoil around Thailand, Burma, Viet Nam, Cambodia, Laos, etc.

The Minister of the Interior was there in full uniform. He corresponds to our Secretary of Defense. He made a pretty speech about the ability of the Thais to protect their borders. A reporter for the U.S. press asked, "You say the reason Thailand is strong is because of the religion and your king, who heads it, and all your Buddhist monks. How do you feel so confident? After all, the surrounding nations are all Buddhists, also."

The minister said calmly and soberly, "Our monks are stronger than their monks."

My objection to the design of the secondary lining continued, but to no avail. I would only stay a couple of months after my contract date. There were two other reasons for this. One was that M&E unilaterally cancelled my insurance under the Longshoreman's Act because of the cost, which was contrary to my contract.

The other was that Peter got sucked in by a proposal of a former employee of mine who thought, if we set up an office in Manila, we could get more work in Asia and so forth. They sent him out to do this and also made him vice president in charge of the area, including Bangkok. He is a good man for what I used him for in Trinidad, on rates and finance. This man is now in the financial section at M&E, where he belongs. Anyway, he got no new business, and I am not sure if he will ever get it.

On my way home from Bangkok the last time, I met the Minister of Finance for Singapore at a meeting of the ADB in Vienna, and he had never heard of M&E. Singapore being small as it is, I don't believe our man could have been very busy there, except for staying at the Shangri-la Hotel. I may have mentioned before that he had been toying with a joint venture in Bangladesh and thought about paying $50,000 graft. Anyway, I went home.

It was not long after that that George Tozer and Peter got me back as a consultant in some management problems, shop drawings, etc., which I had studied twenty years previously. Peter paid me again for a new study, and we found things had not changed much.

Then, in 1980, I did some work for an engineering firm in New York (Forensic Engineering), same as expert testimony perhaps. Hazen & Sawyer, a prestigious firm in the New York area, had some work overseas. Also, they had a similar contract to ours in New York City, just across Jamaica Bay from our job in Rockaway.

A sheet-metal man working for the HVAC contractor had fallen from a ladder because of contact on the back of his neck with a bare wire which was energized. He claimed injuries and sued the city, in addition to the contractors. He could not sue the HVAC contractor because he was covered by Workman's Compensation.

He also sued the consulting engineer, Hazen & Sawyer. By a legal deal, the city and three of the contractors had been left out of the suit, so it only left the electrical contractor and Hazen & Sawyer, their lawyer, and employees. The story was that the electrical contractor was waiting for some motors for louvers.

The louver operation motors had been originally specified as non-explosion proof, which was an error of the consulting engineer and a change order was written to pay the contractor for getting new explosion-proof motors. The electricians had connected the line from the motors to the control panel, which left the line energized and left the ends of the cable just wrapped with insulating tape.

They should have been disconnected at the block at the control center. This should have been done even under the safety code used by Local 3 itself. The current was not enough to harm or electrocute anyone, but it merely startled this man, so he fell off the ladder. Of course, he should not have been on such a ladder alone anyway.

I wondered if, because the engineers had corrected their error which delayed the insulation, they might have more of a duty to check to insulation. Perhaps not. It still would be the duty of the contractor to proceed with good practice and also the union safety code. It appeared to me that the court would hold the electrical contractor liable and clear the engineer of liability.

The lawyer was not too optimistic and felt that Hazen & Sawyer would lose their deductible portion of the professional liability insurance that they carried. It did not get to court, because the injured man offered to settle with the electrical contractor and the engineer at a figure that, in each case, came under the deductible under the insurance policies. The engineer had to accept this, because if he went to court and the decision went against the engineer for more than the deductible, the insurance company would not have to pay the engineer anything, so Hazen & Sawyer lost their deductible in the settlement which amount used up pretty much what they had spent on my work and the lawyer's work.

It seems unfair, but the insurance companies are not in business to pay anything. When you are paying your premiums, they like you. If you have a loss, they treat you like a criminal, with private detectives, etc., So, I piled up thousands of words on contract regulations and so forth all for nothing. Of course, I got my $500/day and expenses.

During this period, August 1978, I had several other calls. M&E on some DEW Line negotiations and some work for the General Accounting Office on Lock and Dam #26 on the Mississippi. This was a dam that had been built during the TVA times, and there had been plenty of trouble with it. They wanted to put a 1,200-foot lock in there, and of course, this was greatly opposed by the railroads. They hired a consulting engineer to set up a method that the existing 600′ lock could be repaired just the way it was without building a new lock.

The Corps of Engineers had hired Raymond to make a study, and the study showed they had to build a new lock. When I looked it over and talked to the GAO, it appeared to me they should go ahead and do what the Corps of Engineers suggested and not to waste money on trying to put tiebacks in through all these wooden piles that no one knew exactly where they were, anyway. The tiebacks were recommended by the engineers hired by the railroads.

But we never took the job, because when we started to think about it, we had clients already—the Transportation Department of the State of Illinois was one, and the other was the Corps of Engineers. For the $50,000 or so we might have gotten from this deal, it didn't seem that we should get mixed up between either of them. Although I gave my opinion to the GSA that they should really go with the Raymond design, and later I believe they did.

Then, I had some talk with a lawyer about some contract in Maryland about a suit that I didn't go on, and I worked on an icing problem for another engineering firm on Lake Superior power plant intake. I kept fairly busy.

♦ ♦ ♦

# CHAPTER 9

## Egypt

In 1980, I started talking with CDM about some work in Manila or Singapore. M&E found out and called me and got me to go to Egypt in November, 1980, on a joint venture with CH2M Hill in Alexandria. This was a job for the Alexandria General Organization for Sewage and Drainage (AGOSD). CDM was still in Alexandria, but M&E and CH2M Hill got the rest of the work, mainly because CDM had set up a master plan that called for discharge into the Mediterranean.

M&E got the job basically because they had proposed, in a review of the CDM reports on wastewater collection, treatment, and disposal, that land application would be a superior method to the CDM idea for two outfalls into the sea.

The AGOSD was headed by Chairman Mahmoud who was a sort of a political engineer, more political than engineer, and they had some capable PhDs on the board. Whereas in Alexandria, the client, Chairman Mahmoud, was not happy with CDM, it was a little different from Bangkok where CDM had the job because Marini had gone to school with Pracha. However, Alexandria was another matter, and M&E was, at least as long as I was there, cooperating with CDM in protecting each other against some of the tactics of the Chairman.

But as we find in many developing countries (President Mubarak once said, "Egypt is not a developing country. It is a backward country."), there was difficulty getting decisions, and when these people get degrees, they wear three-piece suits and do no work.

When I arrived in November 1980, M&E and its local partner, Engineering Consulting Group (ECG) had temporary offices set up in the EI Haram Hotel. This was on the cornice facing the sea. The M&E people lived there, as well as some of the ECG people, and they had offices there.

M&E was in a 50/50 joint venture with CH2M Hill of Portland, Oregon). Incidentally, this was a firm who had been hired by Trintoplan in Trinidad for the work on the water projects, for which M&E had done the preliminaries, and for which we had written the terms of reference for the final design. We had, of course, made a proposal on this work in 1982, and we were told we had the best proposal and the lowest price, but the Prime Minister wished to give it to the local people. These people didn't do five percent of the work, and M&E would have had more local people on the job.

That same year, we had made a proposal to the National Science Foundation to make further study on our Marble Point Report in Antarctica and had been given a reply that the work was to go to Bechtel, because they had proposed to spend more time on the ice and more money, but Bechtel bid a million dollars and our price was $500,000. We had formed a joint venture for that job. We had three principal investigators: 1) I knew the territory and had surveyed and dug holes, etc.; 2) the engineer who placed the nuclear plant at McMurdo, which was to be moved to Marble Point; and 3) the former Navy commander who had laid out the present McMurdo strip on the ice. So, naturally, we would have to spend less time gathering information than Bechtel. At any rate, they made a ridiculously high-cost estimate and the job at Marble Point was never done. Each Naval commander at McMurdo since has recommended the development of Marble Point but to no avail.

CH2M Hill had furnished the principal representative at Alexandria. M&E furnished me as its representative. Les Wierson is a fine fellow, but CH2M Hill is a bit of a PR-type company. We had to furnish résumés of each of our key people. It was interesting to compare Les Wierson's CV with mine. The list of my many years in Trinidad and Tobago included such items as work on preliminary design and development and terms of reference for final design on a comprehensive water plan.

But Wierson's water project résumé was written up as if he had been in direct charge of each project, North Oroouche Dam, Caroni Pumped Storage, Arima Dam, Caroni River, pipelines, treatment plants, pumping stations, and so forth. He had never set foot in Trinidad, but he told me, as president of the International group, he had been in charge.

Peter Gianacakes, president of M&E's parent company, Research Cottrell, hired an Egyptian immigrant architect from New Jersey, Ibraham El Sammak, who had left Egypt during the Nasser years, probably a commendable move. However, in overseas operations, it is not fit practice to send an ex-patriate back to his former country at, of course, a higher rate of U.S. salary than his colleagues in the local firms would be getting. In this case, there was friction plus the fact that he was lazy, did no work, and, of course, as an architect, was naturally deficient in management ability.

The contract was for a 50/50 joint venture, AGOSD and USAID, and had no sponsor for the joint venture. I don't like joint ventures without a sponsor. It was hammered out, and M&E, after plenty of errors, tasks were set up. MRP (Master Plan Review) was not a review but a complete development of a master plan.

TPP (top-priority projects): some of these were already under design and already, in fact, under construction, under CDM supervision. The Smouha Tunnel project was one of the tasks, but no one knew what this was at all.

When I arrived in Alexandria in late November, Elsammak had gone back to the States. I telexed from Boston concerning gaps in the contract and estimated costs, but he wired back that all these things would be subject to further negotiations. After much searching by joint venture people at Alexandria and the ECG people, they had no idea where or what this Smouha project was and what we were supposed to do about it. Neither the client nor co-adventurer seemed to wish me to look into this, but man months and monetary value had been assigned.

What I found after the detailed search of the Smouha area was a German contractor working on a subcontract for Arab Construction Company, a large Egyptian construction firm. The German had completed shafts for two tunnels under the Mahmoudia Canal. Each tunnel is about seventy-five meters long and was designed to act as the inverted syphon for sewage by gravity or to be used as utilidors for force mains, each carrying a couple of 48" pipes for the treatment plant onto the lake.

This German knew his business. He was about to break out from his first shaft and start the shield and was going to hand muck it and then line the tunnel with reinforced concrete rings. Somehow, the tunnel contract was on the way. I never did find out who had started the idea. However, we could consider it in our later planning.

I requested help from our client and our subcontractors to get plans for the existing sewage system. After several days, I was sent a young engineer, Ali, with a large roll of drawings, and we set forth on a reconnaissance. We found that the drawings were of the Municipal Water System and interesting, but not useful to our particular purpose at that time.

The nucleus of the joint venture at Elharam, besides Elsammak, the U.S. personnel, was Saxton, a CH2M Hill man who was acting as project director but listed on the table of organization as Manager of Design. Davis of M&E was listed on the chart as manager of master plan review. Ross of CH2M Hill, Office Manager, and Gus Pantazi, a clerk of CH2M Hill who was to seek out housing for the group who were to come.

Egyptian partners consisted of Sami Abdelkawi, Chairman of ECG, a fine engineer and administrator; Salah, the office man; Gamal Albatrawi, chief engineer who had a decoration from the Russians for building the high dam in Aswan; and a young engineer, Ali; plus Tarik, a young accountant and engineer; and

Michel, an architect helping Gus. Sammy's brother, Medhat, was a sort of general factotum. The only vehicles were furnished by AGOSD and were few and far between.

One of the businesses at hand was the setting up of a job office. A site had already been selected with the connivance of the chairman, and the developer was building a structure, an apartment house, several stories of which would be modified with temporary walls, etc. to be suitable for offices and later rehabilitated for apartments. There was much argument about this, and I thought then, if all this argument was going on about this matter, it did not bode very well for the other decisions that would have to be made during the contract.

Saxton went to a meeting for the choice of sites. Fortunately, he had not even seen the site, drawings, or the status of the construction, but he made a passionate plea, and the chairman agreed. Of course, he got one apartment built for himself.

During the stay in late 1980, there had been bad feelings between Elsammak and Davis: Elsammak, not in line with any prescribed manner of management, chewed Davis out roughly in front of Saxton. Saxton and Davis bickered a lot, also. Then, the backup people in Boston did as all backup people do when you're overseas and did not furnish proper equipment and material. Thereupon, Ross, the office manager, telexed Boston, complaining and suggesting they get their act together.

He was, of course, correct. However, Bill Durham, an ex-Army Corps of Engineers colonel who was designated in charge of the joint venture, sent back a really rough telex complaining about the Ross telex. I telexed Durham that he was hysterical and insensitive to the overseas situation and that, if the people in the overseas site got hysterical, that was understandable, but, in his position, he should keep cool.

These actions by Durham did not make Sami happy, and he was going to get the local board, his firm, and Elsammak, and so forth to vote Durham out of the joint venture. I told Sami this would be bad for the joint venture, and it couldn't be done anyway. I did not tell him, but I knew the joint venture setup would not allow for such action. Anyway, all calmed down, and Les sent a really class telex from Portland, which pleased me and made me feel that we had a good leader in Les Wierson.

My notes for this date, Dec. 14, 1984, read as follows:

> *The field man must be calm and cool for the client and the local people, but he has a right to get excited and even perhaps hysterical toward his Stateside backup people. These people should not answer with a snarl, as this is worse than the usual procedure of no answer at all. The field staff must learn to suffer and learn to accept delay in getting answers.*

Sometimes, we succeeded in having the backup people in the home office use the technique of at least acknowledging questions by sending back, "We will get on this as soon as possible." Andy Paton set up a procedure on our Arctic work, which operated in the best manner.

There, whatever the site man requested (even gold-plated jock straps) would be sent to him, remembering that these site people are at the end of the line. If we found out later that the request was frivolous, we would take disciplinary action, but in the meantime, there would be no second-guessing.

At this time, USAID and the client brought around a new consultant, Boyle & Young, to work right in the client's office in helping manage and monitor the client. At CDM, M&E, and CH2M Hill, this additional force could have corrupted the situation further, but at the time, I wrote a critique for the joint venture, and I proposed we cooperate with CDM, Boyle & Young, USAID, and AGOSD, to make the best of the situation. This cooperation worked, and when the client tried to divide us, we consulted among ourselves and put up a united front. It worked out, and there was no great duplication.

Boyle are engineers and Young are accountants. I remember one case when the client had Boyle & Young's manager, Joe Sewell, a good Texan engineer, doing work on a matter that should have involved CDM, concerning an interim pumping station that was required on the CDM project. Sewell and I listened to the chairman and his henchmen and left.

We went over together to the CDM office and discussed the matter with CDM's manager, George Armes. I drew up some sketches of the temporary pumping station, and CDM completed them, then Boyle recommended the project to the client. The basis of this deal was that CDM had designed a permanent pumping station in the area, but the client had fouled up and did not own the land, so the pipeline already being laid could not reach the pumping station.

The discharge from the permanent pumping station was to go temporarily into an irrigation drain and then out to Aboukir Bay on the west. CDM had designed it so the discharge of the sewage would not get into the irrigation system (there were crops like cauliflower in these fields) but would be carried under a highway by pipe into the drainage part of the irrigation farther down.

Anyhow, for the temporary hookup for CDM's job, the joint venture and Boyle did not join. Further, the client in the long run did not fund the temporary pumping station, and when the pipes were laid as far as the land taking, AGOSD directed the contractors to let the sewage flow by gravity over the irrigation ditches to find its way to the drain in the sea.

Later, when I was working with CDM in Jamaica, Al Weitz, who had worked for George Armes in the Alex area, told me what had happened. So, years later, the question is still not properly solved.

This Master Plan Review, review of the Top-Priority Projects, and the Basis of Design Reports started in early 1981. Davis had the Master Plan Review, and I had the review of the Top-Priority Projects. Saxton was supposed to be the Manager of Design. We found quickly that Walt could not handle this work, and we exchanged jobs. Walt became coordinator, and I, design manager. There were still Top-Priority Projects on my plate.

I had a fine young engineer, Larry Jaworsky of our M&E's Chicago office. He was there only a short time, but he really did a great job on these projects and helped me also in the start of such design, which was feasible at that time. Building of the offices proceeded at a very slow pace. The building was cold and dark and dirty and noisy with construction, debris, etc. In fact, there was no telephone or heat in the offices for eighteen months and very little light.

Maggie and I were fortunate. We went to the Consulate General to register as U.S. citizens, of course, and found there a good list of houses. Houses selected by Gus and his Egyptian assistants and, in some cases, already under lease by the joint venture, were in most cases not suitable. One great problem was the electrical current. Some only had 10-amp services. Gus did not understand enough to check this.

The joint venture contract allowed for rehabilitation and/or installation of heating, hot water, air conditioners, refrigerators, etc. It never froze in Alexandria, but for days and nights of the winter, the temperature stays 38-42° F. Of course, in the summer it may be 95°, but before the stuff came from the U.S., we were pretty cold in the winter of 1982.

We found a villa surrounded by a high wall that also enclosed two really ferocious dogs, Roy and Tosca, German Shepherds. There were two floors. The owner lived upstairs. She was the daughter of Spahi, who had started the textile industry in Egypt, spinning the excellent long-fiber Egyptian cotton. The old man was, of course, a Muslim, and he had four wives. Like a good Muslim, he had to provide each one of them equal benefits, so he built four villas around the city. I understand he used to stay two days at each house.

When Nasser nationalized the industry, the government turned the house over to the Russians. It was a Russian consulate for a while. Nadia, Spahi's daughter (her mother was Jewish and now lives with Nadia's brother in the U.S.) married an Egyptian gynecologist who had interned in the U.S.

Their two sons, Yacque and Kareem, were both born in the U.S. and could be U.S. citizens at age twenty-one, if they wished. Upon their return to Egypt, Nadia and her husband had divorced. She had the boys, but she worried, when they became ten years old, the father could take them away. What happened later on that score, I do not know. Nadia carried on quite a battle with the government for the house.

Telephones in our section of the house still had the number of the Russian Consulate. Of course, while we were there, Anwar Sadat had kicked out most of the Russians, so we used the phones. One evening, I answered the door, and two very proper and dignified Egyptians were there to tell me they would be taking me and my group on a tour of the Egyptian wine industry. They thought I was the Russian Trade Commissioner.

Although Egypt is a Muslim country and alcohol is not universally used, they do make some wine and some rum that is not very great. These Egyptian wine makers used English. They could not speak Russian, but they knew the Russians would be speaking English. Russians, probably because they have so many languages in their own country and given the nearness of other countries, are facile in learning English and other languages.

We found, in Egypt, not only Nadia and her sons, of course, spoke good English, but that English is spoken fluently by many educated people. Our friends at the Yacht Club in Egypt, the Louxors, seemed easily to change from Arabic to Greek to English to German to French to Italian, without missing a beat. At any rate, Nadia's house had 100-amp services on each floor, so the new U.S. electrical equipment worked okay once it arrived.

The villa that Gus had rented for Les Wierson was a beautiful place but had no heat or air conditioning, and the owner had stipulated in the lease that no modifications could be made. Les and his family were cold in the winter and hot in the summer. We were fortunate: we had a fireplace, but it was a project to get wood. There is not much of a forestry products industry in that country. However, it was pleasant to have the fire when we wished.

Our house had no heating or cooling. Nadia had one air conditioner and heater upstairs. Of course, there was none in our place until our equipment arrived, via slow ship and through the long, stupid delays in Egyptian customs. We had a gas stove (propane) in the kitchen. It was very small, and the oven was propped up with a couple of bricks I stole from a building on my way to the office. The refrigerator was small, and the door had to be closed by kicking it shut and to make sure by pushing a chair up against it.

There was an electric water heater in the kitchen, an electric water heater in one bathroom, and one gas water heater in the second bathroom. But the gas operated from a propane gas tank. The trick there was to light off the gas heater early in the morning and start a trickle of water in the tub. This would heat the room a bit, then, after performing ablutions, shaving, taking a shower, I left at 6:30 a.m. My wife would use the other bathroom, where the only heat was the water heater. The kitchen got a bit of heat from the gas stove.

All the other joint venture people except me used various filters and boiling of water for drinking. My wife boiled and filtered her water. I had checked the Siouf water plant, which came into our house, and had no worries about it. I drank the water directly from the tap, and my wife and all the other joint venture people had diarrhea all the time, but not I. This ailment was a subject of cocktail party chatter. The other water treatment plant at Mamoura, I would not drink even if the water were boiled.

In speaking the language in any foreign country, if you need help, try a pharmacy. Generally, somebody there will speak some English. Once, when I was in Thailand, I brushed up on a phrase book and went walking to the drug store to get an eye cup for my wife. The pharmacist couldn't understand what I said.

Finally, when I made some gestures, he said, "Oh, you mean eye cup!" So, that was the end of that.

When I was in Trinidad, for example, the cocktail party chatter would be about servants and food prices, etc. In Alexandria, it would be of diarrhea. I remember the chief engineer for collections, a German, Hans Clausen, telling me almost proudly that he finally got the Pharoah's Revenge.

I think my health was good in Egypt because of my regimen. I ate like an Egyptian peasant. My breakfast was oil, vinegar, garlic, and fool (fool is a sort of Egyptian bean, like a kidney bean, and the farmers eat lots of this in the morning); for lunch, a bit of fruit. I cooked the fool up every morning with oil, garlic, and vinegar in the frying pan and drank carcadee (carcadee is an infusion the flowers from the Upper Nile), kind of a red fluid, something like not-too-tart cranberry juice; some people put a lot of sugar with it, but I like it without any sugar.

At home, I drank rum and water or that lousy wine. After we got our shipments, we had good refrigeration, washing machine, a good stove, and air conditioning, etc. but the power was not always on, so it was somewhat like Medieval England. As Churchill once said, "People lived there in grime and cold and dark."

That was true in Alexandria, both in the home and in the office. The power supply shut off a lot the time, and we appreciated the fireplace. I always had a flashlight in my pocket, so I could find the battery lanterns and the candles, in case the power failed, which it did frequently. One of our Egyptian engineers fell down the pitch-dark stairs of the office and was severely hurt.

We lived right on the tram line; this was one of the few things in Alexandria that worked. This tram line put me to sleep at night, and it started again at 5:00 a.m., so I woke up with it, also. We had to sometimes use it: just step on, pay a few pennies, and get off when you wanted to. It was very safe. One day, an Egyptian man goosed an Egyptian girl and was dragged off by the other passengers. Only the police saved him from being beaten to death.

There was no street crime in Alexandria and no fear of assault, except by the autos. These seem to be operated by people who think they are the only ones on the road, and they appear to deliberately try to run you down. A woman can walk untouched down any dark street without fear. Some spoken offerings sometimes, but touch, no. Justice here is quick and brutal.

My wife would not complain about her Muslim kitchen maid robbing a few bills from her purse. The girl probably would have been severely beaten. In Bangkok, the police would have been just as harsh. While we were in Egypt, a man was convicted of murdering his wife. When she disappeared, he confessed to the crime (police torture I suppose). Just before the execution, his wife reappeared. She had decided to visit some relatives living on the upper Nile.

When you live and work in foreign countries, especially where the language is not English, and especially where the local culture is really exotic, there almost always is a group of expatriates (I don't like that word anyway): foreigners, embassy officials, consultants, and professors, etc.; British, Canadian, American, German, Dutch, French, etc., who tend to gather together, plus a few of the local military, doctors, engineers, etc. who have been educated abroad. We have found this in the West Indies, Thailand, and Egypt.

When Maggie arrived with me at Alexandria, we, as I said before, reported to the consulate and registered and looked up the housing set up. The Consulate General, McClelland, called us at our hotel that same night and invited us to dinner. Mac had been an officer in the U.S. Navy during the war, and all the guests were Egyptian Navy captains, four of them, and their wives. One was still on active duty. His wife was an artist. Another was retired, an attorney. And then, there was Hassan Louxor, whose wife ran the USIS Library. It was a pleasant, small dinner, and we saw all these people many times in Alexandria and still correspond with some of them.

McClelland left shortly after and was replaced by Jim Bahti, who was a great art lover. He and his wife were close to us during our tour. Hassan brought us to the Yacht Club of Egypt and put us up for membership. We met his sisters; one of them, Zizi, was a bachelor girl (unusual in that society). We keep up correspondence with her and Hassan. Zizi worked at the Misr Bank, a government-owned bank. (Misr is the Arabic word for Egypt.) The country allowed U.S. dollar accounts, so we kept some U.S. cash there that we had electronically transferred from the States; it was at 17-1/2% interest.

Hassan operated an oil tanker set up at French Harbor. Tankers came in an out and unloaded and loaded oil. Egypt was not then entirely free of need for imported oil. By now, she is probably self-sufficient. Hassan regulated the flow. It was educational to be in his office and hear him talk on the radio, giving orders and discussing arrivals and departures with skippers of ships in the harbor and some out of view. He speaks many languages, and he could switch frequently from one to the other.

His father had been an admiral of the Egyptian fleet. Like the Danes, the Egyptians seem to have mainly submarines as their capital ships. Hassan was also a skilled sailor of many-sized sailing ships and in sailing competitions all over the world. He was a great friend of the sailor King of Norway, and I told him about my nephew, Bill Gleason, who also was a good sailor with a king as a friend: the King of Thailand.

Friday is the Muslim Sabbath, so we took Friday and Saturday off but worked on Sunday. On Fridays, we had a routine. We went to a grocery store. Our visits there were a regular thing. The storekeeper was a good Muslim, so he did not carry alcohol, but would send out to get it for us.

It was just a sort of ceremony. The manager would have a coffee or soft drink with us (not much English here), and it was hardly a supermarket, though it had green vegetables, and in season, the vegetables and fruit are fine. The cabbages grow so big, we used to buy one for two families.

The fool and carkedee had to be bought in the souk, but there was a meat market next door to this shop, and the proprietor had ground-up hamburger. Seldom did we have any other kind of meat, as it was all tough. Then, we went to the duty-free shop, where we would sometimes, courtesy of the USAID, buy Scotch and cigarettes. I was happy with the local rum anyway. Duty-free

Scotch was $3. The local stores sold it for $17 and $18. Sometimes, they would put duty-free stamps on them. So much for the black market.

The rest of the day was spent at the Yacht Club. The bar did not open until all the Sabbath prayers were said by the men in a large tent next door; it was not large enough to prevent a huge overflow on the adjacent property. Friday lunch at the club was always good.

Alexandria shrimp were huge and tasty. We could sail with Hassan or visit some of the yachts that called in there. The club was rather posh, and the harbor of Alexandria is a pleasant sight, especially from a distance. There would be a fair-sized group of us most Fridays.

Zizi had a sister, Amelia, whose husband was a brigadier with a bristly mustache who had been stationed with the Egyptian Embassy in Bonn, Germany. He was the official secretary of the Yacht Club of Egypt. The group would include Egyptians, sometimes Gamal Hafez, who was chairman of PetroJet, a pipeline company. He and his people operated this like the best of the U.S. pipeliners.

Admiral Khattig, one of the PetroJet's directors, was often there, but usually he was at the Greek Yacht Club next door. He was a great oarsman, and the Greeks had a rowing team. Saxton and his wife and others from CH2M Hill, Joe Sewell of Boyle, and of course, Zizi, Hassan, and quite often the English engineer who worked for an air conditioning firm. Hassan's wife was seldom there. Hassan had a real roving eye. Although they had two handsome early teenage sons, they did not operate together very often.

But the great thing about the Yacht Club of Egypt was the yachts and their crews who sailed them out of there. A British yachtsman had a yacht, *Moon Maiden,* with two masts but no sails. It was a tops'l schooner with very fancy rooms and baths and galleys better than any kitchen in Alexandria. The skipper was a former Royal Navy British captain and had his wife aboard.

Zizi was madly in love with this captain. However, his wife kept a good eye on him. We were on board several times and entertained the captain at our house. The ship was on her way to the Red Sea to pick up her rich British owners. She had been chartered several times for lots of money.

Hassan met one of the girls on board, a stewardess, and volunteered to take the ship through the canal for the captain one day. He left with the yacht and traveled to the island on the Red Sea. As Hassan's wife said, "He is off having a good time." By the time we left Alexandria, I believe she had found herself a young Egyptian engineer, a little younger than she. She was a very attractive lady.

Then, there was a charming French couple who sailed into port in a handsome aluminum sloop they had built themselves. Both of them were good mechanics, so the welding was impeccable. They were on a honeymoon cruise of five years around the world. I remember Joe Sewell, saying to them, "Oh, that costs lots of money." They gave him a pitying look, like JP Morgan saying, "If you have to ask what it costs to own a yacht, you can't afford it."

Then there was an American who sailed in with his crew and got paid off right away. Some of the joint-venture people got conned into taking some vacation time and sailing with this captain to Cyprus. They practically got shanghaied once they got on board. They had to work, so the men got blistered feet and sunburned, while the wives had to do all the kitchen work and cooking. Once at sea, the captain told them to work or it would be mutiny, so it was not a pleasant vacation for them.

We met a group of Swedes, all members of a Swedish skydiving club, who sailed in from Singapore. Only one of them had ever sailed before, and he knew no navigation. He had bought the sloop in Singapore, and three of them had sailed it across the Indian Ocean and came in via the Red Sea. There were no engines, and they tacked back and forth, learning navigation as they went along.

Hassan and I suggested they go back through the French canal system, hook an outboard motor on, and stay away from Biscay Bay, which was very rough at that season of the year. They did not wish to pay the tolls to go through France, so they took off, and I guess they made it.

Hassan set up an international regatta. He called up to our house one night when the consul general and his wife, Anita, where there with several others. Hassan got Jim to set up the U.S. corporation. It was quite an affair. The only U.S. entry was a young lieutenant submariner who sailed a Laser at the races that were held in Aboukir Bay (where, years ago, Lord Nelson had defeated the French fleet).

There was a big dinner at the club. Each team was there, flags were flown, and various national anthems were played. Egypt, Greece, Italy, France, Germany, and the U.S. participated. All the ambassadors and consuls were there, too.

Maggie used the club to entertain several of our friends visiting us. I seldom could get away to lunch. We had a couple of drivers, brothers named Said and Mimi. Their father, Ahmed, ran the corner newsstand. Each day on my way to work, I walked up the hill, and Ahmed would greet me with a big handshake. When Said came on Friday morning to drive us to the market and club, he would bring some falafel and aish that his poor mother used to cook for us.

Maggie, of course, joined the American Women's Club. Non-Egyptian women could belong, including French, Dutch, and British. It was a bridge group and made life pleasant. There was also a painting class, where the teacher was an Egyptian, and Maggie learned a lot from her. The Sporting Club had a pool and golf course, where, except for Fridays, when hordes of Japanese descended from Cairo, there were hardly any players on the course. Consulate and joint-venture people made up the whole population on Saturday.

A friend of Maggie's, an English woman married to an Egyptian engineer, and her husband called on us. Abdel Nasif had been prominent in the construction and operation of the tram line. Adbel and Evelyn used to arrive just after my return from golf on Saturdays, and we had a bridge game. We also used to get together at their beach house in Agamy, a very pleasant place. Maggie learned a lot painting over there, and she painted a dozen or so beautiful pictures, for which she was offered considerable money to sell them, but she wouldn't.

There were two other Egyptian engineers: old Rader, who was using a walker, and Ishmail, representative for International Nickel. Jack Rader, Ishmail Abdel, and George Armes used to meet sometimes and talk engineering stuff. Of course, we were all in our seventies. Abdel died, Rader became bedridden, George Armes went home to die, and Jack and I figured Maggie's health would be better away from Alexandria.

We still correspond with Zizi and Hassan and, in the U.S., with the Sunds and the Prestons and few others. On the professional side, Jack was glad to get away. M&E set up a replacement for him as Design Manager and then replacement for Construction Manager. He was glad to get out of a frustrating situation. It was hard for him to take the project seriously. It was a profitable venture for the two firms and their employees, and for the ECG and their employees, it was fine, but after the Master Plan Review was getting done, I couldn't believe completion would come in my lifetime, or at least not in the time I might be able to spend at useful work.

In about July 1981, Jaworsky asked me if I thought the job would ever be finished. I said that I would not live long enough to see the first gallon of treated effluent go out to sea or to proper land disposal. He was about forty years younger than I am, so he might have been around to see it. At any rate, despite no telephone, the telex facilities far away, our offices being built around our heads, and the auxiliary actions by client, the job progressed—not that anything got done, but a master plan was developed.

This master plan ran into trouble when Davis of M&E and Suhr of CH2M Hill tangled. I could not convince some of these people that the joint venture meant a joint effort. Davis gave sixty days' notice. I told Les he should accept it right now and let Davis go home. Les's answer was to appoint Suhr as Chief of Master Plan

Review and let Davis stay on for sixty days. I objected, but Les persevered.

The chairman got into the act and, in mid-1981, announced to the public and the press that AGOSD was going to give up the idea of an outfall and would be putting the effluent on land and make the desert bloom like a rose. This was not an acceptable position to USAID. They said they did not wish to go into the agriculture business.

Around this time, we found that these sewage treatment plants, which we were supposed to improve, were unable to handle sewage anyway, and sewage pumped into the plant by the Abu Soliman Pumping Station was being discharged into the lake untreated. AGOSD had a PhD lady chemist who told me the sewers could not be treated by biological methods because of the great mineral content of the industrial wastes pushed into the system.

The consultant for the industries involved, Dr. Ludvig, was insistent and adamant that the industries had no responsibilities to pretreat their waste, and because of the benefits of the industry to the economy, the government should treat it. I criticized the USAID people for leading, in a way, the consultants to think that they could find a way to treat the material and to think that one plant was treating sewage.

McAleer of the USAID told me he was not expert at all on wastewater treatment, but that in the negotiation phase, he had taken the supposed best sanitary engineer in the world to the existing plant, and no one said anything about the material going unchanged through the plant.

He also said that USAID would not enter into agriculture and would not finance the land disposal idea here. I told him we were talking about three million people in Alexandria and the local area and their welfare, and the proposed spending of $9 million over ten years to help these people could only be compared with the money spent each year to provide arms to three million people in Israel ($9.2 million a year). And for this project, of course, it would be spending for real benefits. Weapons are, of course, obsolete in a year or so.

Davis was still hanging about; he was at a party at the Syrian Club. The client, the joint venture, and ECG were all there. The Arabs did not drink alcohol, so the chairman was sober but Davis was drinking, and he started to talk to the chairman about the master plan. He told him he wanted to talk to him in private, to tell him the truth about the new plan. When I came in the next morning, the ECG people were in an uproar. Gamal and Sami were going to have Davis deported immediately. I listened to the story, as Les was away and I was in charge.

Davis was not really repentant; however, I told Sami that I would take care of the matter, no deportation. I told Davis that, at the next meeting of the steering committee, of which he was a member, I wanted him to get up and praise and defend the proposed master plan in front of the chairman.

I told him, if he did not do this and show he was loyal to the joint venture, I would get him blacklisted as a proper engineer and employee, and he would never have another job or position of trust. He had to agree. I did not go to the meeting. I told Sami, if Davis fouled up, to get him out of the meeting. Jaworski was there. He told me that Davis did a good job of defending the master plan review. Jaworski said he thought that Davis was a little nervous. I knew damn well why he was nervous.

Then, I gave a farewell party at my house, and Sami and all of us made him feel good. He had a little Arab girl he had picked up somewhere when he'd been in Egypt before. She was sort of a housekeeper for him. At the going-away party, he sat next to my landlady, Nadia, and spent all his time telling her that he wanted to marry this little girl. He brought her back to the States with him, but I understand he let her visa run out, and she was sent back to Egypt.

We were still struggling with the office. One new floor was completed, and we wished to put the collection group, headed by Hans Clausen, to this next floor, the third floor. This group was then on the second floor. As I have said, the building was designed as an apartment house, so the use as a design office was somewhat difficult. For an engineer in the collections group, we had picked up an engineer of CH2M Hill who had been in Saudi Arabia and was working in Alexandria without a contract, but was due to return to the US to pick up a new contract and return with his wife and family. This man, Brady, had spent most of his time as a civil servant in Public Works Departments in the U.S.; he'd spent most of his time in Alexandria playing with his pocket calculator.

He did not believe he should move upstairs. I told him he had to, and he said no more to me. Then, Hans came to me when they were moving and said Brady had threatened him. Told him, if he had to move upstairs, he would quit and go home, because he did not really have a contract. He was a little bit terrified.

Brady was a huge, red-bearded Irishman. I called Brady in and told him he had threatened our man. When I said our man, I meant the joint-venture man, not M&E or CH2M Hill. He said he had not really threatened him, but he had made the statement.

I told him he had better get packed and that he was all done. Les was due back in a day or two, and Brady asked if he could speak to Les. I told him he could speak to him, but he was leaving anyway.

When Les came back, Brady saw him, and Les said that Brady had apologized and wanted a second chance. I told Les he had not taken my advice about Davis, and if he gave Brady another chance, he would come back with his wife and children and the family cats, and when he fouled up again, as of course he would, it would be a mess. I know that being ruthless on overseas work is necessary: seven days a week, twenty-four hours a day in close quarters. You must be heartless in such circumstances. Les agreed, and we solved that problem.

I said, before that, our relations with CDM, Boyle & Young were always good. When we had the problem with the temporary pumping station, I had written a memo saying we should recommend the contractor working on the CDM project at

Raselsuda should be terminated, to keep the irrigation fields free of untreated sewage, and the job should be shut down. The contractor should also be better instructed in the matter of deep sheeting and laying of the flexible pipes. Davis misunderstood this and made some remarks that CDM was not properly monitoring the work.

LD Taylor and CDM engineers had a useless argument with Davis on this. The chairman delighted in the friction among the Americans and met me one evening when I was coming back to our office. He appeared to be waiting for me.

He said, "I just heard that CDM has been giving you such a bad time, you were going to leave the country."

I had not heard about the Davis & Taylor incident. I told him, "CDM and I get along fine. George Armes is one of my dearest friends." Then, I called George and his people and got it all straightened out.

One of the many roadblocks was that the client had designed, by themselves, a sewage collection system on a project of the government buildings in the Smouha area. When the contractor started it, we had already found that the AGOSD engineer had used an incorrect elevation, and some of the pipes would be corning out of the ground with the slopes they had designed. We got that straight by showing them how to reduce the slope.

Then, in March or late February 1981, we were called in to look at a concrete encasement for a siphon containing a 48" force main for Abu Soliman Pumping Station to the Eastern plant. The 48" pipe was twelve millimeters-thick steel, except for some bends and rises which were only six millimeters, and these were corroded. They looked like lace curtains, and the sewage leaked out of the pipe and into the concrete utilidor, and then into the Mahmoudia Canal. The leak was below the Siouf Water Treatment Plant, but the flow finally discharged into the Mansoura Water Treatment Plant.

This plant had trouble anyway, but this additional load was so great as to make disinfection impossible. In order to fool the public, the AGOSD was only pumping at night, so people could not see the sewage running into the canal. I recommended a quick solution:

put a pump into the encasement, lay an 8″ steel pipe across the road, and cover it with gravel. It was a cobblestone pavement. Run this pipe into a manhole, which would take it back to the pumping stations. This could have cost an extra pumping, but it would also keep the sewage out of the canal.

AGOSD would not do that, but after some time they laid a pipe by gravity back to the manhole. This took a long time, and in the meantime, they had all that raw sewage getting into the canal. I already had talked to Gamal Hafez of PetroJet and developed a bypass plan, which would shut off the flow on each side of the canal; with a few bends and some river weights, we could build a bypass around the leaky pipe. This pipe could then be repaired. The price for this was around a $150,000, but even our ECG people thought the price was excessive, and the client would not act at all.

At Christmas 1983, I got a letter from my replacement in Egypt, Bonnie Claussen, showing a photo of the bypass pipeline and on top of the ground, which, of course, would shut off the navigation of the canal. Up to then, the 48″ pipe had not been repaired. What was too expensive in March 1981 was okay at the same price as a less valuable bypassing in 1984. I heard, by Christmas 1984, things were finally moving toward getting some real work done and straightening out the Alexandria wastewater situation. It had been four years already, and no substantial of work has been done. By 1986, in fact, there was still no action.

We left Egypt in January 1982, in time to be back in Hingham for Maggie's birthday. We had had a great time in Rome in October 1981, at a conference for the Comité de l'Arctique, and while we were there, Sadat had been assassinated. When we returned the day before his funeral, we found things very quiet. We went to a small exercise with the Consulate people and some other U.S. nationals to discuss the situation of evacuation and so forth, but nothing serious happened.

We stopped in Switzerland on the way home and then Amsterdam. Maggie broke her shoulder in a fall in our hotel room in Amsterdam. We let the children think for a short time that it happened while skiing in Wengen. Then, we went to Tucson for a short re-retirement.

CDM's Jim O'Rourke, who had been vice president of M&E and was now with CDM, called from Boston and asked us to go to Poughkeepsie, New York on an industrial waste project for IBM. We had a beautiful apartment in Fishkill, New York, about the best we ever had. The job was fine for me.

Great to have a client like IBM: they knew what they wanted, they expected you to know how to get it done for them, and they were willing to pay fair value. They did not expect to waste any money. It was delightful not to have to pinch pennies or to have stupid policies just floating around.

It was okay for me, but for Maggie, it was pretty desolate in the backwoods New York State town. You know when you are overseas, as I have said, there is always a small society of people who are in the same boat as you are, and life is pleasant. But in a town in the U.S. where people live their own lives and pay no attention to strangers, it may not be pleasant.

This IBM job was unique in that, while the project was to build a plant for the treatment of industrial wastes, the groundwater was so contaminated with various organics, the water pumped from the foundation excavation had to be treated before it was discharged into the river. IBM is very pollution-control minded, and with the USEPA and New York State people bugging them, they lean over backwards.

This site had been used by several other companies. One was an ammunition plant during a couple of the wars. There was much waste dumped all over, and there was plenty of contaminated soil. In one area of Poughkeepsie, some wells were being contaminated by things that had not been put there by IBM. Without charge to the house owners, IBM had laid water pipes direct to the area from the city water system and corrected a problem that was not even theirs.

The IBM contract read, if any contaminated soil was found, the contractor would report it. IBM would test it, and outside contractors would remove the material and haul it to a waste disposal facility. This material would be handled with double containment trucks. However, if the outside skin was broken, there would still be a continuous skin to prevent the discharge on the highway.

The construction water from the excavation was pumped into the tank trucks and then the fluid run through activated carbon filters, to remove the trichloroethylene contamination, which is the residues from the solvents. Residue was collected from the sludge pits and hauled to the authorized waste dump by special contractors.

The contractor for the excavation job had set up a well point system to lower the water. This produced a great deal of fluid and it would be very costly to treat it all, because of the huge amount of water to be treated before discharge. I proposed that the contractor drive close sheeting then muck it out in the wet, placing the excavated material where it would drain and take out the organics. Then, when the earth was all out, we would tremie a plug of concrete in the bottom for the seal. This would allow the removal of only the water in the pit, which would be pumped into the old sludge bed, treated with activated carbon, and tested to check the parts per billion (ppb) of the contaminant; it contained 200 ppb, and only 50 ppb were allowed for discharge to the river. We attained less than 5 ppb.

However, the contractor was not really a general contractor, and I had objected to giving him the job. He did not know what to do and drove some poor wooden sheet piles, then could not properly brace the sheets to excavate. So, we got a regular type contractor who came, drove steel sheeting, mucked, tremied the seal, and pumped what little fluid was in the hole to the sludge pits, treated it, and then the scavengers hauled the residue to the dump.

It is not always best to give a contract to a low bidder. The money you save for the low bid may delay completion of the job, and the low bidder may lose money and delay the job. The owner can save money in the long run via getting the job done early by selecting perhaps the third bid.

I left this project for the Jamaica assignment. I left the job in Gene Penzimer's hands, who had worked with me and M&E in Greenland.

♦♦♦

*Alexandria, Egypt*

*Egypt-2*

*Egypt-3*

*Egypt-4*

*Egypt-5*

*Egypt-6*

♦ ♦ ♦

# CHAPTER 10

## Jamaica

The Jamaica job was a little different from the usual consultant relationship. CDM sent six of us to work in Jamaica on a contract with the Jamaica National Investment Company. This had been set up by Prime Minister Seaga, something like how the U.S. set up the RFC (Reconstruction Finance Corporation). In Jamaica, they had just formed a National Water Commission (NWC). Up to then, there were several authorities responsible for the delivery of water. The National Water Authority operated all over the island except in Kingston.

The Kingston Water Commission supplied water to the metropolitan area of Kingston, and the KSAC (Kingston and St. Andrews Corp) oversaw over ninety little water utilities in the environment of Kingston. The two adjacent parishes of St. Thomas and St. Catherine's had almost a hundred other little systems, with customers from 75 to 7,500.

There were plans to make the National Water Commission the utility that would handle all the water and sewage for the whole country. This is a tremendous job, 4,500 square miles. I don't know of any other utilities that have two million people spread around such an area and one authority to operate it. In Massachusetts, for example, there are almost six million people over 8,000 square miles, and there are many water authorities.

In the area of the South Shore, for example, wells are the town supply. Hingham has a private company. Weymouth has some town sources and some purchased from MDC. The attempts of the

Jamaicans to set up headquarters in Kingston to operate the whole country was a commendable but difficult task.

KSAC, for example, has ninety-plus small utilities (they only send bills to 3,000 customers). Parish deficient in each case of only minor drought with maintenance, leak repair, treatment of deficiencies, pump rate down, and huge problems in chlorine supply and handling of great plagues.

The next thing is the lack of equipment, parts, etc. In 1983, CDM helped to get an IBID loan, which should help in getting personnel and spare parts, but it will be an uphill fight. I had been in Jamaica in 1969, and nothing had been done since then toward increasing either the storage or the supply.

A deep well-drilling program had been started, but the operators were amateur hydrogeologists and inexperienced drillers who produced a failed program. The program was expensive and a fiasco. The minister in charge, in reference to the preliminary report leading up to our project, starting in March 1983, wrote in July of 1982, "I don't want any more damn reports. Just tell us what to do, and we will do it."

We sent down six "not experts," as the local columnist wrote, not consultants but specialists. Each was to be the counterpart of an officer in the NWC. The NWC had a board headed up by a political appointee, the chairman. He was a big betting-pool operator, a bookie, and he had been in business as a snow cone vendor. He was very rich, and also the major man in the town where he had a regular castle. He presided over the town of theirs like a sultan. He had, naturally, been a large-sized contributor to Seaga's campaign.

The Managing Director was a former official in the old national water authority who knew nothing about the business. CDM sent as a counterpart a former administrator of the Tampa utility. CDM sent a counterpart for the finance director, another refugee. The counterpart for the director of training was a retired regular Army Corps of Engineers officer, who had been for several years head of the Wastewater Division in Denver. The NWC man with him had returned to Jamaica from an assignment in Canada, and there was some resentment among the people because of this. CDM sent one of their vice presidents from California, who was surplus at the time, to be counterpart to the Director of Engineering, who was a Jamaican who had studied at the M&E office in Boston.

The Director of Operations was formerly the Director of Engineering for NWA. He was an old-type civil servant. He had two offices: Manager of Operations for the Kingston Water Commission and Manager of Operations for the rural districts.

CDM sent me to be counterpart to the Manager of the Kingston metropolitan area, and a former CDM man whom I knew in Egypt was going to be with the rural distribution. Pennant was manager for Kingston, and Mitchell was manager for rural. They were good men, but they had problems. Each of us would sit in the same office as our Jamaican counterparts and try to help.

As shown above, there was no counterpart for the director of operations. This left Lawson, the Jamaican, in a different position from the others, and he tried to run the whole show before the board. When we arrived, the reservoirs in Kingston were dry. The only available surface water was the inflow of the surrounding rivers into these two reservoirs. The few wells operating were overpumping it and picking up salts.

CDM's report on Jamaica in June 1982 made reference to the development of a Kingston strategy, a method to be devised to ration the water, the water between rainy seasons. This was needed in Kingston rather than the rural districts, because although there is lots of rain on the windward side of the Blue Mountains, Kingston is on the lee side and does not get much rain.

The discussion in 1969 concerned itself with the possibilities of tunnels and so forth to bring water through the mountains to Kingston. This had been a suggestion for many years. I found a 1930 report that discussed in detail the same basic idea that we reached there: shut the pumps down when the reservoirs were full, and when the service level dropped, use the pumps. However, not enough wells were in service, and it would be required to lock off the water in certain areas at specific hours to preserve the water in the reservoirs. Pennant had set up a schedule to do this rationing in 1982-83.

I had not seen his schedule, but when I set up the schedule for March 1983-March 1984, it coincided with Pennant's. But in spite of Pennant's pleas and arguments, the management lost on the board, and the politicians would not allow Pennant to set up his rationing station. They wanted to give the people, in late 1982, all the water they could for the holidays, so by March 1983, the reservoirs were dry. Trucks were running around, and tanks were being looked at for schools, etc. It was an unnecessary emergency.

The pumps were pumping too much chloride. We had only been there a few days when the whole bunch, including us, were called to a meeting before the Prime Minister. The PM knew his party had refused to let Pennant ration when nothing was done in December 1982, which was why this disaster had just arrived.

Lawson said they had to give the people water for the holidays, and then later, if the water were left in the reservoirs, it would only evaporate anyway. Anyway, it finally rained in June and again in September; however, the reservoirs did not fill completely. In December 1983, I submitted a new rationing plan, tying the levels to the reservoir against the water use and, by rationing, to be sure there was some water all the time.

Here again, by this time, Pennant had been transferred out. The same charade continued. The management would not let Daley, who was by now the whole management of Kingston, start locking the waterfalls. This was for political reasons. The PM has said he was under criticism for some other matter.

When he was told, he said, "I can't cut the water before the holidays. I'm in enough trouble now."

We did, however, get them finally to select to cut out water here and there and to use the tank trucks, so it was not as bad as the year before. But if we had not battled for this, it would have been just as bad.

In the meantime, the new wells, which were supposed to be fast-tracked, all failed, and the rivers pumping in Halls Green that picked up water below the temperatures outflow got all fouled up, because the intake was not properly designed. The same engineers who designed that facility had designed a new facility at Castleton, and for political reasons, that project was put under contract.

There were two things wrong with this. One was, when there was plenty of water in the Halls Green site, the Hermitage Reservoir was overflowing. When there was plenty of water at Castleton, the dam would be overflowing. But in spite of my growls, the Castleton project had the go ahead to redesign it, because the design was the same as Halls Green; the pumps were plugged up, so the intake had to be redesigned. Halls Green had to be repaired.

The Directors were, as I said, not competent or knowledgeable. They sat in their ivory tower and held meetings—cut down overtime, cut out temporary help, etc. Our fearless leader, the counterpart for the Manager and Director, got the board to issue a rule: no more temporary help. Then, we were short a man at the intakes. These intakes were cut in eight little streams up the Point Hills, and the water was picked up and transmitted to one line to help fill Hermitage Dam, which was also supplied by the Whitewater River, and also to supply the water treatment plant at Seaview.

Another line to supply the water treatment plant was at Constant Spring. FIDCO (Forest Industries Development Company), a government agency for lumbering, got some kind of permission from the Water Commission to build more roads up in the catchment area. But the Forestry Department had been planting trees over the years. The Commissioner and the Director had never been in the area and evidently could not even read a map.

FIDCO cut right across the mountainside to Mount Creesy, with side hill cuts and fills, so that, when there was rain to carry water to the intakes, the sand and gravel would wash into the streams and the intakes would get plugged up. This increased the workload of the Water Commission people.

They always had done some work, of course not so much with the FIDCO waste coming down. It would depress them. When their intakes got plugged, they hired some of the local farmers and loggers and hunters, etc. who lived up in these hills, to clean the intakes.

This was work for these people who had a right of access to these intakes. It was not like 5th Avenue, but these mountain men were used to it anyway. Then, the order came out: no temporary help. Our peerless leader said that NWC had hundreds of men, and we should take some out of the Kingston wastewater crews to do this work.

He, like all the others, had never seen these places and had no knowledge of local customs there or of mountain people anywhere. One of the city boys could not do the work, and he would not do the work, plus it would take all day for them to travel, and they might even get lost in the woods. The other thing was that the local people would resent it and might even resort to violence.

Jones and I and Daley told Egbert Jones, the superintendent there, to hire these men as temporary help, and if necessary, we would pay for them personally. We got it done, and the directors never complained.

Then, they ran out of chlorine. I had set up a chlorine program based on the AWWA recommendation and my opinion of the local scene, and I had given it to Lawson, plus a copy to Rudolph Brown, the chief of chlorination operations. He had nothing to do with ordering the chlorine, though.

Brown had requisitions for parts of his chlorination units out for about two years. He wasn't getting the parts, but being versatile in skills, he had been keeping the chlorination going. He had all kinds of Mickey Mouse hookups. Even when the chlorine ran out this time, Lawson and the Board did not wish to invest money into proper supply. The ship got broken down and did not arrive. Brown had made hookups of ton containers, of which there were fair supply, and hooked them up to the 150-pound cylinder plants, which helped out.

One of our tasks was to set up a schedule of comparative tests at the proper location, to determine the chlorine residuals. CDM, at my request, spent some money for which we never got reimbursed to buy more equipment. The program was set up for each zone superintendent to check these places every day and report. It sounded good, but the superintendents dedicated people who had some instruction to use this and who had some time to use their own Volkswagen to haul pipe, because there were no trucks.

We couldn't expect to be diligent in such tasks when some of the old pipelines were corroded, so the chlorine would not get there but would be used in the process, so no residual. Chester Rodney, for one, called me up just after we had gotten the truck started. He couldn't find any residual at the airport or at the end of the line at Fort Royal. I went out with him and found this very true.

A lot of people check for contamination once a week, and the water was okay, except there was no residual with the chlorine to prove it. These pipes and many others had been badly tuburculated over the years. Maurice Jones had gone to Florida and gotten a price from a polypig operator, but no one pigged these lines, because there was no money.

It is hard in a case like that for even a dedicated superintendent to believe he is not wasting his time, checking for something he cannot find. Our counterparts for Director of Engineering were, like Daley, opposed to temporary help, although he, like the rest, never went into the field. He complained to me one day that his house got water most of the time from the Forest Hills pump, and it was salty.

I told him, which he should have known anyway, that he would have to get used to salty water, but if he needed water for drinking, he could come to my house and fill some jugs.

My counterpart was pleased and complimentary on my choice of a house that was in an area where the water could not be locked off except for local repairs. This was not an exact science, only a guess, and of course, I had only been there ten days before I made the deal. However, we had no telephone for six months. There were a couple of places where we could have moved to, but then, we would have to rise at 4:00 a.m. to fill our bathtubs because of the rationing.

This counterpart business did not always work. During the development of a nationwide idea, the directors got a man to be national head for transport. He was just an ordinary clerk by trade who, I was told by Rudolph Brown and others, owned a share in a large garage and repair shop that did business with the authority. He was stuck in an office at the newly developed headquarters.

However, his actions did not endear him to the drivers and mechanics. One day, there was a bit of a mob scene. The truth is always hard to find. Mr. Knight was escorted out of his office and off the lot, but there was no violence. He complained and said he did not dare to go back unless he had armed guards.

Our peerless leader panicked and wrote a memo to the chairman, saying they should discharge thirty-nine drivers and machinists, etc. This was mass punishment. Two of the machinists I knew wouldn't harm a fly, but they attacked the chairman for standing behind us.

We tried to get the machinists back to duty, but some people told the chairman that they continued to keep the thirty-nine people unemployed. Then, the personnel people and the union and lawyers found out the procedure was to bring charges, have an investigation, and keep the men on 75% pay until it was settled.

Believe it or not, that went on for over four months: thirty-nine men getting paid three-quarters pay for the work they should be doing, farmed out to local shops including the NNN, allegedly partly owned by Mr. Knight. All charges were finally dropped. Knight would not testify, so it is, in my opinion, CDM's fault in part of it.

One of our tasks was to work with police, fire, and the Office of Disaster Preparedness to set up plans for the use of water in the event of disaster, e.g., fire, earthquake, hurricanes, etc. My first attempt at this was to inquire into the use of sea water to fight fires and the docks and waterfront. I found that the fire boat was inoperative. There were two tugs that had firefighting equipment on board, but no one knew how to operate them.

The fire department did not wish to use salt water, anyway. Commander Harvey of the Office of Disaster Preparedness, a Royal Navy ex-commander (he put me up at the Yacht Club for membership) said that the fire department used foam that was not compatible with salt water.

I told him and he agreed, of course, that there was suitable foam, although it would be a little more expensive. And such salt water systems were all corroded. I said that all the ships that sailed the sea pumped salt water, and all ports I had seen had salt water pumps and pipes. It would be ridiculous to fight fire with precious fresh water all the way from the reservoir.

Later, Mr. McDonald, Harvey's boss in the Office of Disaster Preparedness, said that he talked with the PM, and they were going to repair the fire boat. I know they painted but did nothing else.

To go on with the Office of Disaster, it was not on a fault, but the pipeline from the dam to the water treatment plants crossed at least one fault, and there were no repair pipes available to repair such a break. Daley told me just before I left that he had finally ordered some pipe and fittings to repair that, if necessary.

I worked out with the electric people, Jamaica Public Service, their response time, if the transmission lines went down. The supply hospital's standby generators were priority: JPS was the first, hospitals second, and water commission would be about the same.

NWC could not operate with JBS, and JBS would have to get water for their boilers. Both the local JPS at Hunts Point was on the shoreline, and so was the refinery and the oil tanks. To conform to our aim to help the hospitals, it appeared we must get the pump in shape and also get in on 415 volts rather than 2300 volts, because JPS was going to modify his transmission so that Forest Hills pumps should all be placed like all the other water commission installations. This was done before 1985. The Public Service financed some of the work. After that, NWC would be all alone, and JPS would have no 2300-volt service to offer. I don't know what happened with that.

It appeared that a 1963 hurricane had filled the reservoir with about five to six million gallons of mud, which needed to be taken care of. The dredge still floats in the pool. This was the second dredge; the first one did not hold up.

When I flew over in 1969, I saw they were pumping upstream so that the next rain would bring the water back again. However, in 1983, I found they had pumped upstream but let it dry out. It took the shoveling trucks to move the material. This also filled its tunnel into a constant spring that required a temporary line and a large mucking out process.

♦♦♦

# EPILOGUE ON ENGINEERING

Is the engineer an insensitive materialist whose activities are hostile to humanistic values? Is engineering an expression of man's most elemental impulses or his most sublime aspirations? Has engineering brought our civilization to the brink of ruin or made it possible?

Engineering, like medicine, predated science. Engineers study the sciences and apply them to problems of practical interest, most often through creative design. But what does it mean to be an engineer? And how do engineers think about what they do?

It is often said that engineers have narrow interests, that they are indifferent to human relations, to the social sciences, and to the arts. There may be some truth to this, but it is far from the whole story. Creative technical work can provide satisfaction much like painting, writing, or composing music. Engineering is an art; no art can be fully learned.

Engineers use existing knowledge, ancient and modern experience, imagination, and intuition to design, build, and maintain the engines of our civilization. Scientists, engineers, and technologists depend on one another; yet the public often accepts the benefits without understanding the fundamentals. When trade-offs appear, the blame is placed on the producers and on engineering, itself.

Take fire. The technology of making and carrying fire was among the first developments that distinguished man from other animals; without it, man could not be man. Yet today, as in all ages past, we do not fully control it; thousands of lives and vast property are lost to it every year. We must have fire, no matter how primitive or how sophisticated our technology. This pattern, of an essential benefit paired with persistent risk, runs through all of technology.

We live in a complex technological society that often does not handle its own technology very well. Failures are frequently blamed on machines; more often, they arise from inadequate training, poor maintenance, or simple lack of understanding.

That may be one of the most important things I have learned: technology magnifies human qualities. It does not correct them. We like to believe that better machines will solve our problems. Sometimes, they do. More often, they simply make our strengths more effective and our weaknesses more dangerous.

Do engineers qualify to shape public policy? They design and build the systems that support modern life; yet politics dominates decision-making. Engineers respond to public demands but are often blamed for the consequences of those demands.

There is a need for understanding on all sides. Engineers benefit from exposure to the humanities; non-engineers must also understand something of science and technology. A society that depends on complex systems cannot function if most of its members cannot understand even their operation.

Education remains a central issue. Colleges may produce specialists with little appreciation for broader thought, while many graduates struggle to understand even simple technical material. At the same time, highly educated professionals in other fields may lack basic practical knowledge.

The development of the whole person, the training of individuals for service to the community, may be the highest problem of life. The betterment of conditions throughout the world does not depend solely on scientific knowledge, but on the fulfillment of human traditions and ideals.

Man's greatest task for anyone's success lies in the exploration of all things - physical, mental and moral.

Men may not be naturally aggressive, selfish, or greedy; yet whatever the cause, the struggle for survival persists at every level. From the lowest-paid laborer to the highest-paid entrepreneur, each is driven in some measure to lead, to compete, and to secure his place.

When I was at MIT in the early 1930s, we were required to take courses in business law and accounting. It was early familiarization, much like training in the Army; not to make us professionals in law or accounting, but to learn enough to know when we needed help. As Royster said, we have passed the point where one man can command all human knowledge. Yet education should still provide people with the basic skills needed for life and an understanding of the workings of the civilization in which they live.

I have known a PhD accountant who lived for a year with improperly installed plumbing that made his apartment unlivable, yet he did nothing about it. An eminent psychiatrist boasted of being unable to balance a checkbook. These are not trivial matters; they reflect a deeper lack of practical understanding in a technological world.

Engineering practice itself is shaped by economic pressures. Fees are negotiated to a minimum, and something must yield. Details are left to contractors; specifications become subject to interpretation; disputes inevitably arise.

I recall advising young engineers who are working for a contractor, when bidding a job, to examine three things: general conditions, the contract documents, and inconsistencies between drawings and specifications. If there are loopholes, they will be found. A contractor who has bid his life's blood on a job is prepared to take advantage of them.

When I ran a job by myself in 1933, Louie Perini gave me only one instruction: get paid for what you do. If you move ten yards of rock, do not try to get paid for twelve, and surely do not settle for eight.

Technology itself is neither inherently good nor bad; it has always had both constructive and destructive aspects. Iron gave man tools to build and to cultivate and became weapons to kill and to torture. Environmental concerns illustrate this dual nature.

The few years man has been present on this Earth, he has built various civilizations. Our archaeologists dig them up today: cities buried in their own midden heaps, and irrigation canals silted in because of lack of maintenance. All past civilizations have failed because of lack of maintenance. Eventually, to maintain our present civilization, we must care for far more than was required, centuries ago.

The problem is not technology itself but how it is used. Many people rely on complex systems with little understanding of how they work.

Society, for the most part, with its majority who has at least graduated from high school and perhaps even attended college, often cannot read a simple manual of operation.

It is not the fault of the technologists, the scientists, and engineers. There does not seem to be enough technologically literate civilians. Before the motor vehicle, those who drove horses understood far more about the care and feeding of their animals than their descendants now understand about their automobiles.

Vermont Royster wrote that education should prepare people for life and understanding civilization.

The production line has been blamed for despair, yet it is necessary. Our industries and schools seem unable to train people to run and repair the engines of our civilization. We have the right ideas, but the execution falls short. This gap between design and use is dangerous.

The competence of automobile driving tests is questionable. My father once said that to be a successful driver you must carry the continuing thought that every other person on the highway is a complete idiot and must be expected to do the wrong thing.

There is also the matter of ethics and of luck. We rely on both more than we care to admit. Ethics must guide decisions where rules and specifications fall short; yet even the most careful design and the most honest effort are sometimes overtaken by chance. Good fortune is often mistaken for good judgment, and bad luck for failure, but the two are not the same.

Technology has always had two facets, benign and threatening.

Tyndall wrote that life existed long before man. Some species, perhaps man himself, may eventually be endangered, obsolete, and finally extinct.

Calvin Rutstrum, in his *Chips from a Wilderness Log,* says, "Nature is patient. The billions of people on the Earth will pass with the flick of Mother Nature's finger." Other billions, more or less, will take their place, recycle the elements of past generations, and create new generations. It is as though a single individual is of no more significance than a grain of sand upon the ocean shores.

Rutstrum recognizes that man did not weave the web of life; he is merely a strand in it.

Whatever he does to the web, he does to himself. Wilderness or not, the preservation of the world still exists so we can enjoy it today plus save it for coming generations. But if we don't leave our descendants a habitable, life-affirming world, we will deserve to be forgotten, and the willingness to forget will be our eternal death.

Technology cannot be the only culprit. In Dostoevsky's *Notes from the Underground,* the central character rejects comfort, order, and rational systems, insisting instead on the unpredictability of human desire. He values the process of striving more than the attainment of any final condition, and he resists every attempt to define or prescribe his needs. There is something deeply human in that resistance.

I have often thought about Sisyphus, who was condemned to push a boulder up a hill for eternity, only to have it roll down each time he reaches the summit. Had he been an engineer, he would have chucked the boulder at the top and sat down to rest.

Yet, we cannot turn away.

We cannot escape what we have made, nor should we try. The question is not whether technology will continue to advance, but whether we will advance with it. We have the knowledge, the tools, and the experience; what remains uncertain is our judgment, our discipline, and our willingness to understand both the power and the limits of what we create.

*John and Maggie Scheuren*
*"Every man has his price. Mine is my wife and my children."*

♦ ♦ ♦

# John J. Scheuren, Jr. Chronology

Born in Massachusetts in 1909, John Scheuren lived on every continent of the Earth except Australia. He served as an officer in Army Air Corps in World War II in North Africa, Italy, Southern France, and the Mediterranean.

Fellow - Explorers Club – Emeritus

Marine Technology Society – Emeritus

International Association of Permafrostologists – Expert Bi-Polar

American Society of Civil Engineers – Fellow

Arctic Institute of North America – Emeritus

Registered Professional Engineer in NY, NJ, MA, CT, and RI

MIT (BA Civil Engineering, BA Mechanical Engineering) 1931

1930s: Superintendent and Engineer for Perini Corporation in the 1930s for construction of highways and bridges on the Boston-Worcester Turnpike; Boston Post Road; Grand Central Parkway in NY; Quabbin Dam and Dike, highways in New Hampshire; Cambridge to Concord Highway and bridges, and road and bridges on Podunk Road (Charlton to East Brookfield); dam at New Hartford, CT; widening of the Cape Cod Canal; Newburyport Turnpike; road, bridges and tunnel at Worcester; Lincoln Tunnel in NY and NJ; Tuscarora Tunnel on Pennsylvania Turnpike; road and bridges in Chelmsford, MA. Returned from Caribbean in 1972 to work for Perini North River Associates on a quarter of a billion-dollar project in NYC for sewerage work, drilled in caissons, slurry walls, dredging, connection to tunnels, outfalls, and intakes.

1940s: General Superintendent and Manager of Construction with C.S. Foreman Company of Kansas City, Missouri in the 1940s for large cross-country pipelines for natural gas, crude oil and other products and their appurtenances such as pumping and compressor stations in Kansas, Missouri, Nebraska, and Illinois. Also, water treatment plant piping in Missouri; Syracuse Air Base; Romulus Combat Equipment Depot, Target Range and Railways at West point; railroad work in NY and NJ; and the War Emergency Pipeline ("Big Inch") from Texas to New Jersey (400 miles of 20" to 24" pipe and 13 major river crossings).

1950s: Metcalf & Eddy as Architect-Engineer Manager and Overseas Manager on work in Arctic, Greenland, Iceland, Northwest Territories in Canada including exploration, reconnaissance, survey, design and construction of airfields, aircraft control and warning stations, Ballistic Missile Early Warning Systems, docks, and pipelines (1952 – 1961). For this work he received a Citation from Chief of Engineers US Army – Patriotic Civilian Award. "His accomplishments reflected great credit to the Corps of Engineers and to the United States Army and enabled the Corps to carry out its military construction mission in the National Defense. (1952-1961)"

1957-58: In 1957 he led a group of engineers and scientists to study the feasibility of building an airfield and appurtenances on snow-free land in Antarctica. During this study they built a small airstrip at Marble Point, the first in the Antarctic. As part of the exploration in the Marble Point area, a river noted in the area is now called the Scheuren Stream.

1960s: In 1962, Scheuren participated in "Operation Water Hole," a terrain analysis study for the US Air Force for locating and preliminary design for alternate landing areas for aircraft and space vehicles.

Vice President and Director of Projects for M&E in the West Indies. Projects included sewerage, water supply and housing developments in Trinidad and Tobago, irrigation in St. Lucia, highway in Guyana, water supply in Antigua and an airport in Grenada (1964-1972)

1970s: 1972 with M&E for construction management on rebuilding and extension of large Water Pollution Control Facility in Rockaway, New York. In 1976 went to Bangkok, Thailand as Vice

President and Project Director for the Bangkok Municipal Waterworks on construction of tunnels, reservoirs, and treatment plants. In 1979 was M&E's representative on a joint venture in Alexandria, Egypt on improvements to sewerage and drainage systems.

1980s: In 1982, he went to Jamaica, WI for reorganization of the Jamaica National Water Commission. After his last "retirement" in 1984, he consulted on such diverse projects as an overview of the famous great slide at Fort Peck Dam in 1938 and of the failure of the shuttle Challenger. His collaboration with M.R. Montgomery was noted in Montgomery's book on the Fort Peck disaster.

Publications include Boston Society of Civil Engineers (Construction Division Best Paper) "Arctic Construction" 1962.

Additional Selected Publications, published and unpublished:

"Engineering in a Temperate Tropical Area"

"Notes on Construction under Adverse Weather Conditions"

"Industrial Wastes"

"How to Plan and Pay for the Safe and Adequate Highways We Need"

"Black Power Demonstrations"

"The Engineer as Entrepreneur"

"Fertilizer from Sewage"

"Engineers as Professionals"

"Treatment of Rum Wastes in Stabilization Ponds"

"Pelicans, Fish and Sewage Treatment in Trinidad"

"Industrial Use for Waste Treatment Plant Effluent"

"Water Management and the Inuit"

## Invention: Drink-o-meter

# John J. Scheuren Jr., 85; worked around world as civil engineer

By M. R. Montgomery
GLOBE STAFF

John J. Scheuren Jr., 85, of Hingham, died Wednesday in South Shore Hospital in Weymouth after a career in civil engineering that took him from Egypt to the Arctic, from Jamaica to the Antarctic.

Like so many civil engineers of his generation (he graduated from MIT in 1931), he could build anything – gas pipelines, dams, sewage and water systems, airfields and high rises. He could sink caissons and clear minefields, a skill he learned during World War II in the US Army Corps of Engineers.

Mr. Scheuren never lost his sense of humor about the military. He was always proud to say that as an MIT graduate with a dozen years of experience on major construction projects, including the Winsor Dam that holds back the Quabbin Reservoir and the Pittsburg, N.H., Dam on the Connecticut River, the first thing the US Army did was send him to cooks and bakers school. He ended up in Egypt during the North African Campaign, clearing mines, rebuilding railroads, creating airfields and water systems, and dealing with thousands of Italian troops who had surrendered in the early Libyan campaigns.

He developed a special affinity for, and knowledge of, the problems of constructing facilities in the most difficult conditions – the permafrost, snowcap and weather of the polar regions. From 1952 to 1961, he was overseas manager for the Boston engineering firm of Metcalf & Eddy, working on defense projects in Greenland, Iceland, Antarctica and the Canadian Northwest Territories.

Engineers of his era had enormous practical abilities, often taking direct charge of machinery and explosives that would now be in the hands of a specialist. He recalled his work on the widening and lining of the Cape Cod Canal with pride and his usual good humor.

The first plan for widening the canal, he said, would have required enough pumps to drain the Atlantic, and so he broke the job into manageable bites, shorter sections, and widened and rip-rapped the canal behind sheet piling. When the day came to blast out the foundation of the old Bourne Bridge, located directly under the main high-tension electrical line feeding Cape Cod, he placed the charges himself, and remembered with satisfaction that not only did the electricity keep flowing, but the whole excavation slid neatly and harmlessly into the Canal.

JOHN J. SCHEUREN JR.

Mr. Scheuren, among dozens of honors, was a fellow of both The Explorers Club and the American Society of Civil Engineers.

He leaves his wife, Margaret (Burlingham); three sons, Fritz of Virginia, Jeffrey of Colorado and Mark of Hull; three daughters, Marie Williams of Norwood, Margo of Mansfield, and Claire of Arizona; two sisters, Harriet Gleason of Marblehead and Elizabeth LeFavour of Virginia; 12 grandchildren and two great-grandchildren.

The funeral will be tomorrow at 3 p.m. in First Unitarian Church, Cohasset. Burial will be private.

From *The Boston Globe*, Boston, Massachusetts
20 May 1995, Page 71



♦♦♦

# Pioneers of Polar Science, Exploration, and Engineering

The mid-twentieth century marked a transformative era in polar exploration, scientific discovery, and Cold *War–era infrastructure development*. The individuals profiled in *Talking Through My Hard Hat: Chronicle of a 20th-Century Civil Enginee*r represent a remarkable convergence of disciplines: glaciology, aviation, engineering, geophysics, and military logistic, whose combined efforts advanced human understanding of the Arctic and Antarctic, while enabling sustained operations in some of the harshest environments on Earth.

Henri Bader (1907-1998) was an influential Swiss-American glaciologist and Chief Scientist at SIPRE/CRREL. A key figure in the exploration of the Greenland ice sheet, particularly in the Thule region, he played a central role in planning and directing glaciological programs for the International Geophysical Year (1957–1958). His pioneering work on snow physics and metamorphism helped establish modern scientific understanding of snow and ice behavior. Bader Glacier was named in his honor in 1958.

Bernt Balchen (1899–1973), a Norwegian-born aviator who later became an American citizen, was a pioneer of polar aviation and a recipient of the Distinguished Flying Cross. During World War II, he applied his Arctic expertise to support Allied operations over Scandinavia and Northern Europe. Among the earliest to demonstrate the feasibility of sustained aviation in polar regions, he later served as a senior U.S. Air Force leader and consultant on Arctic and aviation projects.

Richard Evelyn Byrd, Jr. (1888–1957) was an American naval officer, pioneering aviator, and one of the most prominent polar explorers of the twentieth century. He led multiple major expeditions that expanded the geographic and scientific understanding of Antarctica and helped establish a lasting U.S. presence there. His achievements included aerial exploration across vast polar regions and the discovery of Mount Sidley, Antarctica's largest dormant volcano.

Anthony A. Cucolo, Sr. (1897–1983) was an industrialist, philanthropist, and civic leader whose life embodied the American immigrant experience. Arriving from Italy at age sixteen with minimal means, he rose from laborer to president of multiple companies. His contracting firm constructed the roadway for the Tappan Zee Bridge, and he played a prominent role in the civic and economic life of Suffern, New York.

George John Dufek (1903–1977) was an American naval officer, aviator, and polar expert whose career spanned World War II, the Korean War, and extensive Antarctic operations. Working first with Admiral Byrd and later as a senior leader of U.S. Antarctic programs, he achieved a historic milestone in 1956 by becoming the first person to land an aircraft at the South Pole.

William Maurice "Doc" Ewing (1906–1974) was a pioneering American geophysicist and oceanographer. His work in marine geophysics and seismic research contributed fundamentally to the emerging understanding of ocean basins and played a significant role in the development of the theory of plate tectonics.

Sir Vivian Ernest Fuchs (1908–1999) was a British scientist and explorer who led the Commonwealth Trans-Antarctic Expedition (1955–1958). Under his leadership, the expedition completed the first successful overland crossing of Antarctica via the South Pole, one of the last great geographic achievements of the heroic age of exploration.

Sir Edmund Percival Hillary (1919–2008), the New Zealand mountaineer who, with Tenzing Norgay, first reached the summit of Mount Everest in 1953, also played a major role in Antarctic exploration. He led the New Zealand component of the 1955–1958 Commonwealth Trans-Antarctic Expedition (TAE), achieving the first overland crossing of Antarctica via the South Pole on January 4, 1958, using modified farm tractors. He was later knighted for his contributions to exploration and public service.

Count Eigil Knuth (1903–1996) was a Danish explorer, archaeologist, cartographer, artist, and one of the foremost authorities on northern Greenland. Over his more than six decades of Arctic fieldwork, he conducted extensive expeditions in Peary Land and other remote regions of Greenland, mapping previously little-known areas and documenting important archaeological evidence of early Inuit cultures. Renowned for his endurance, scholarship, and deep knowledge of Greenland's geography, history, and Indigenous heritage, Knuth became a central figure in twentieth-century Arctic exploration and research.

Fr. Daniel Linehan, S.J. (1904–1987) was a Jesuit priest, geophysicist, and founder of Boston College's Weston Observatory. During U.S. Navy Operation Deep Freeze, he conducted the first seismic measurements of the Antarctic ice sheet, determining a thickness of 8,197 feet at the South Pole. He also served as chaplain aboard the *USS Atka* and celebrated the first Mass at the South Pole. His contributions advanced early understanding of Antarctic ice structure.

Dr. Nagaya (Tomohiro Nagaya, 1900–1962) was a Japanese scientist specializing in snow physics, permafrost, and the behavior of snow and ice in cold-region environments. In John J. Scheuren, Jr.'s recollections of Greenland field operations, Nagaya is described as a highly regarded expert whose work contributed to understanding snow densification, glacial formation, and the structural behavior of snow under load. His insights were frequently referenced in discussions of engineering challenges on the Greenland ice cap, particularly in relation to the mechanics and long-term behavior of compacting snow and ice.

Robert L. Nichols (1904–1995) was an American geologist and explorer whose work significantly advanced knowledge of the geological history of Antarctica and Greenland. Over the course of seventeen expeditions, including five to Antarctica, he conducted extensive field research, discovered and named Tufts Pass, and endured extreme conditions, including a prolonged period of isolation sustained only by seal meat.

Ootah was a legendary Inughuit explorer who accompanied Admiral Peary and Matthew Henson to the North Pole in 1909. He was inducted into the Explorers Club's Society of Forgotten Explorers.

Andrew C. Paton (1909–1988), a partner at Metcalf & Eddy, was a leading engineer in large-scale infrastructure development during the Cold War era. He oversaw the design and construction of major international projects, including Thule Air Base in Greenland, military installations in Spain, and water and sanitation systems in Trinidad, as well as projects in Vietnam, Iceland, Saudi Arabia, and Alaska.

Robert Peary (1856-1920) was an American explorer and officer in the United States Navy who made several expeditions to the Arctic in the late nineteenth and early twentieth centuries. He was long credited as being the discoverer of the geographic North Pole in April 1909, having led the first expedition to have claimed this achievement.

Louis "Louie" Perini (1903–1972) was an American construction executive who transformed the Perini Corporation into one of the largest construction firms in the United States, operating globally across four continents. His company played a major role in large-scale infrastructure and military construction projects. He was also the principal owner of the Boston and Milwaukee Braves baseball franchise from 1945 to 1962.

Capt. Peter Joseph Smenton (c. 1910–after 1957) was a United States Coast Guard officer and polar operations specialist. From

1956 to 1957, he commanded the icebreaker USCGC *Westwind,* playing a key role in Cold War–era polar logistics.

Col. Morton Solomon (dates unknown) served as a district engineer with the U.S. Army Corps of Engineers and was noted for his leadership in construction safety. At Thule Air Base in Greenland, his teams achieved accident rates far below national industry averages. In 1954, the National Safety Council recognized this achievement with a special award, highlighting the importance of safety in remote, large-scale engineering operations.

Lieutenant Commander Henry E. Stephens, CEC, USN, was a Civil Engineer Corps officer who played a key role in construction efforts in Antarctica. Operation Deep Freeze was a U.S. naval operation that supported scientific efforts and established a permanent presence on the continent.

Paul-Émile Victor (1907–1995) was a French ethnologist and polar explorer who became a central figure in twentieth-century polar research. He led a 1936 dog-sled expedition across Greenland and later directed major scientific surveys, including a 1951 study demonstrating that Greenland's ice sheet overlays multiple land masses. Founder of the French Polar Expeditions, he was awarded the Patron's Medal of the Royal Geographical Society.

Derek Wright was a prominent New Zealand National Film Unit (NFU) cameraman known for his extensive filming and photography in Antarctica, particularly during the 1957-1958 Trans-Antarctic Expedition (TAE). He documented Sir Edmund Hillary's journey to the South Pole, capturing historic images of the team and their tractors.

♦ ♦ ♦

# Index

www.ingramcontent.com/pod-product-compliance
Lightning Source LLC
LaVergne TN
LVHW100251210926
848480LV00003B/892